Robert Recorde

Title page of *The Castle of Knowledge*

Robert Recorde

The Life and Times of a Tudor Mathematician

EDITED BY
GARETH ROBERTS AND FENNY SMITH

UNIVERSITY OF WALES PRESS
CARDIFF
2012

Reprinted (paperback), 2013

www.uwp.co.uk

British Library CIP Data
A catalogue record for this book is available from the British Library.

ISBN   978-0-7083-2682-4
e-ISBN   978-0-7083-2517-8

Designed and typeset by Chris Bell, cbdesign
Printed by CPI Antony Rowe, Chippenham, Wiltshire

Contents

CONTENTS

List of illustrations

Unless noted otherwise, illustrations from Recorde's books have been provided courtesy of TGR Renascent Books.

Notes on contributors

JOHN DENNISS is an independent scholar with an extensive knowledge of early English arithmetic texts.

GARETH WYN EVANS (1921–2005) was appointed lecturer in Applied Mathematics at University College Swansea in 1949. He taught at Swansea for thirty-five years, specializing in relativity, quantum mechanics and thermodynamics. He had a strong interest in the history of the development of mathematics, particularly the concept of time.

STEPHEN JOHNSTON is Assistant Keeper at the Museum of the History of Science, University of Oxford. He first met Robert Recorde in the 1980s, as a graduate student working on Elizabethan practical mathematics. His research and teaching still focus on topics in the histories of early modern science, technology and mathematics.

HOWELL A. LLOYD is an Emeritus Professor of History at the University of Hull. His research interests lie in a range of early modern British and European subject areas, and nowadays principally in the history of ideas.

MARGARET PELLING retired in 2009 as Reader in the Social History of Medicine in the History Faculty, University of Oxford. She is now based at the Wellcome Unit at Oxford as a Senior Research Associate. Her research on health, medicine and social conditions focuses primarily on the Tudor and early Stuart periods.

NIA M. W. POWELL is a lecturer in the School of History, Welsh History and Archaeology at Bangor University where she specializes in social, economic and cultural aspects of early modern Wales.

ULRICH REICH lectured at the University of Applied Sciences in Karlsruhe in Germany. His research interests in the history of mathematics include the work of Johann Scheubel (1494–1570) and his link with Recorde.

GARETH ROBERTS is an Emeritus Professor of Education at Bangor University. He has particular interests in the acquisition of mathematical concepts within bilingual contexts.

FENNY SMITH is an independent scholar with knowledge of ancient and medieval numerical notation and arithmetic techniques, and Italian Renaissance algebra.

JACQUELINE STEDALL is a Senior Research Fellow in History of Mathematics at The Queen's College and the Mathematical Institute, Oxford, with a special interest in early modern English mathematics and the development of algebra.

JOHN V. TUCKER is Professor of Computer Science and Head of the School of Physical Sciences at Swansea University. His research uses algebra and logic to understand computing systems. He also has a particular interest in the history of computation, including its social, economic and cultural contexts.

JACK WILLIAMS is an independent scholar with interests in Robert Recorde's biography, the metallurgy of coinage, mercantile arithmetic and unit fractions.

Acknowledgements

IN BRINGING THIS PROJECT to fruition, the editors have enjoyed the support and encouragement of many individuals and organisations. The work has taken its inspiration from the extensive groundwork completed by Jack Williams and his vision to ensure that Robert Recorde's life and accomplishments are fully documented. We have also benefited from the ready advice and constant encouragement of Jacqueline Stedall, and Gordon and Elizabeth Roberts. Without the timely financial sponsorship provided by Bangor University, the British Society for the History of Mathematics, WALMATO and WJEC, the venture could not have been completed. We have also appreciated the support of staff at the Tenby Museum and Art Gallery, who have worked tirelessly to disseminate information about the most famous son of that delightful town. Particular thanks are due to the family of the late Gareth Wyn Evans for permission to reproduce his 1958 lecture as an appendix in this volume. We are also indebted to the individual authors, who have shown endless patience and endured gentle prodding by the editors with good grace. The constructive advice provided by an unnamed reader was greatly appreciated and we wish also to express our thanks for the assistance of the team at the University of Wales Press: Ennis Akpinar, Sarah Lewis, Siân Chapman, Dafydd Jones, Henry Maas and Catrin Harries.

Preface

R OBERT RECORDE has had a 'makeover'. It has been a long process and is still to mature fully as his importance to the development of mathematical thinking becomes ever clearer.

He is most popularly known as the Tudor mathematician who invented the equals sign, but other facets of Recorde's life are also well documented. These include his extensive mathematical publications, his work as a physician, his often turbulent service to the Crown, and his keen Protestantism.[1]

In Wales, the first public acknowledgement of his work took the form of a bust by Owen Thomas erected in 1910, the assumed 400th anniversary of his birth, in St Thomas's Chapel within St Mary's Parish Church in Tenby, Recorde's birthplace. This image of Recorde is based on a portrait subsequently shown to be of someone quite different[2] and there are some inaccuracies in the accompanying stone memorial tablet. Nonetheless, the image and the biographical sketch together provided an iconic reference point and a challenge for further work.

One of the uncertainties yet to be fully resolved is the exact year of Recorde's birth. The issue is teased out further in this volume by Jack Williams, who has undertaken exhaustive research to 'set the Recorde straight'. Without his encyclopaedic knowledge, this book would not have been possible.[3]

The year of Recorde's death is more certain and its 400th anniversary was commemorated in 1958 at a conference in Trinity College, Carmarthen. The event was organized by the Faculty of Education at Aberystwyth, and was supported by the south-west Wales branch of the Mathematical Association. A report on the conference appeared in the *Mathematical Gazette*.[4] It is significant that the impetus for the conference arose from a faculty of education and that it was supported further by an organization concerned with the teaching and learning of mathematics.

By 1964 those within institutions of higher education in Wales involved with the teaching and learning of mathematics (mainly, but not exclusively, as part of teacher training) came together to form an organization still known by the somewhat curious acronym WALMATO.[5] From the outset, WALMATO enjoyed the support of members of Her Majesty's Inspectorate working in Wales and representatives of local education authorities across Wales. As well as providing a network for those involved in mathematics education, WALMATO's main activity centres on an annual conference held at Gregynog, a University of Wales centre situated in mid-Wales. Robert Recorde rapidly became WALMATO's unofficial honorary past-president, sometimes irreverently but affectionately referred to by its members as 'Dai equals'!

At the celebration of its twenty-fifth birthday in 1989, the WALMATO conference included a paper by Alun O. Morris, professor of mathematics at the University College of Wales, Aberystwyth, on 'Robert Recorde and notation in mathematics'. The paper was subsequently included in the proceedings of the conference, *Mathematical Education in Wales: Twenty-five Years of WALMATO*, an internal publication edited by Tom Brissenden and Glyn Johns, mathematics tutors in the Department of Education at the University College of Swansea and the University of Wales College Cardiff, respectively.

At the turn of the millennium, Robert Recorde was celebrated at the University of Wales Swansea when, in February 2000, its Department of Computer Science named a seminar and conference room the Robert Recorde Room. The Department, led by John V. Tucker (one of the contributors to this volume), subsequently commissioned a large slate plaque, the Robert Recorde Memorial, designed by the artist John Howes and carved by the calligrapher Ieuan Rees. This plaque, unveiled in 2001, adorns the entrance to the Robert Recorde Room. Recorde's work and influence was now being recognized outside the confines of the disciplines of mathematics, education and Tudor history.

The 450th anniversary of Recorde's death in 2008 provided the ideal opportunity for WALMATO to act as the initial catalyst to draw together these various strands in a conference devoted exclusively to Recorde's life, his work and his continuing influence. Crucially, interest in the proposal was expressed by other mathematics organizations and, in particular, the British Society for the History of Mathematics (BSHM) readily agreed to partner WALMATO in the conference planning.[6] The conference also benefited from support generously provided by the Welsh Assembly Government and WJEC.

Publication of the conference proceedings had not been the original intention, but the success of the event and the enthusiasm engendered by the realization that a wealth of research was available, coupled with the cooperation of the University of Wales Press, led inexorably to the preparation of this book as an edited version of those proceedings.

The process was further assisted by Gordon and Elizabeth Roberts, who had attended the 2008 conference and had, 'over coffee', shown delegates their personal facsimile copies of Recorde's books. Their subsequent decision, after some gentle cajoling, to make copies available on a commercial basis,[7] has assisted the editors of this volume enormously and has also made copies of the originals readily accessible to interested readers. Gordon and Elizabeth have also augmented their publications with valuable additional information about Recorde and the background to each book.

Finally, during 2010 the town of Tenby celebrated the 500th anniversary of Recorde's birth. The celebrations included a memorial service at St Mary's Church, coinciding with the 800th anniversary of the church itself, and a memorial lecture at the town's Museum and Art Gallery. The latter hosted a series of readings from Recorde's books during the summer holiday period and also staged an exhibition, *2010−1510 = Robert Recorde*, based on the response of over fifty artists to Recorde's life and work.

Notes

1   See, for example, Stephen Johnston, 'Recorde, Robert (*c*.1512–1558)', *Oxford Dictionary of National Biography* (Oxford, 2004), online edn accessed 21.11.09.

2   The portrait is now believed to be Flemish, dated 1631, and is housed at the Department of Pure Mathematics and Mathematical Statistics of the University of Cambridge; see J. W. S. Cassels, 'Is this a Recorde?', *Mathematical Gazette*, 60/411 (March 1976), 59–61. See also chapter 3.

3   See the seminal work by Jack Williams, *Robert Recorde: Tudor Polymath, Expositor and Practitioner of Computation* (London, 2011).

4   W. S. Evans, 'S.W. Wales Mathematical Association: report for the session 1957–1958', *Mathematical Gazette*, 43/343 (February 1959), i–ii.

5   Welsh Association of Lecturers in Mathematics in the Area Training Organization. Following the publication of the McNair Report in 1944, the School of Education of the University of Wales was designated by the Ministry of Education as a single Area Training Organization. See D. Gerwyn Lewis, *The University and the Colleges of Education in Wales 1925–78* (Cardiff, 1980).

6   For a review of the conference, see Gordon Roberts, 'Robert Recorde: his life and times', *BSHM Bulletin*, 24/1 (2009), 40–2.

7   Facsimile copies of Recorde's books are published by TGR Renascent Books. Details are available on the publisher's website, *www.renascentbooks.co.uk*.

Editorial conventions

QUOTATIONS FROM Recorde's texts are reproduced using Recorde's own spellings and punctuations, unless otherwise indicated. In particular, the title of each book corresponds to the spelling on the respective first editions as follows:

The Ground of Artes
The Vrinal of Physick
The Pathway to Knowledg
The Castle of Knowledge
The Whetstone of Witte

Reference to these books in the text or in footnotes is often shortened to *Ground*, *Vrinal*, *Pathway*, *Castle* and *Whetstone* respectively. Full details of Recorde's books are provided in the bibliography at the end of the volume.

Recorde's printers made extensive use of diacritical marks to indicate the omission of the consonant m or n. For example, the title of *The Ground of Artes* appears as *The groūd of artes* on the title page of the first edition. In general we have replaced such marks with the full spellings. Equally, contractions that appear in the original texts are written out in full so that, for example, yᵉ and yᵗ are reproduced as 'the' and 'that' respectively.

Quotations from Recorde's books are generally referenced in-text. Unless indicated otherwise, such references are to the first editions. In most cases, individual pages are referenced using signature conventions such as in (*Ground*, sig. A.iiiᵛ), where the superscript v indicates a left hand page. Square brackets are added to indicate that the signature mark, or part of it, has been omitted from the original, for example (*Ground*, sig. B[.i]). By contrast, page numbers are used to locate references in *The Castle of Knowledge*, in which the numbering is generally reliable, but a reference such as (*Castle*, p. 127 [129]) indicates that the quotation can be located on page 127 of *The Castle of Knowledge*, although the number on that page has been misprinted as 129.

The editors have been greatly assisted by the availability of facsimile copies of Recorde's books published by TGR Renascent Books, Derby, and by advice liberally provided by the proprietors, Gordon and Elizabeth Roberts.

Introduction

GARETH ROBERTS AND FENNY SMITH

A CRITICISM OFTEN LEVELLED at Recorde was that he was not a first-rate mathematician and that his contribution to the development of mathematics was minimal. In 1920 Frank Morley dismissed his mathematical work as being 'hardly memorable'[1] and shades of such criticism are not uncommon even today.

The flaw in such remarks is not that they are incorrect but that they completely miss the point. Recorde himself made no claim that he was pushing forward the frontiers of mathematics. He was, rather, a communicator of mathematical ideas who sought to explore ways in which to make mathematical knowledge and skills available to a wide population. Above all else, he was a mathematics educator, an insight first clearly underlined by Geoffrey Howson in his seminal work in 1982 on the history of mathematics education in England: 'Not only did Recorde teach mathematics, but his writings show clearly – both implicitly and explicitly – that he had also given serious consideration to the problems of learning and teaching mathematics.'[2]

The theme of Recorde as a mathematics educator is common to a number of the contributions to this volume, particularly in relation to his publications on arithmetic (*Ground*), geometry (*Pathway*) and algebra (*Whetstone*). In his description of the essence of the ideal teaching process, Martin Trow could well have been summarizing Recorde's methodology: 'Teaching is not an action but a transaction; not an outcome but a process; not a performance, but an emotional and intellectual connection between teacher and learner.'[3]

This emotional and intellectual connection between teacher and learner is developed by Recorde through the adoption in three of his books of a dialogue style between Scholar and Master that enables him to explore ideas in a leisurely way, revisiting difficult points, introducing elements of wit and humour to enlighten the exposition and, ultimately, at the end of *Whetstone*, to signal his own demise.

The use of the vernacular throughout his books emphasizes the importance Recorde attached to being able to communicate with a wide audience, and his concern with points of vocabulary highlights his attempts, not always successful, to adapt the English language to enable it to express mathematical ideas with precision and clarity.

He emphasizes the need to provide motivation for new mathematical ideas in the learner's mind, to aim for understanding (as opposed to learning by rote), to make wide use of examples to illustrate the application of new ideas, and the need to practise that which has been taught and learned (*Ground*, sig. Y.ii): 'vndoubtedly it is practise and exercise that maketh men prompte and experte in euery kynde of knowlege.'

Howson's analysis of the development of mathematical education in England coincided with the publication of the highly influential Cockcroft Report,[4] the deliberations of a committee of inquiry into the teaching of mathematics in schools in England and Wales. In clear and succinct terms the report set out the principles that underpin good teaching. It is instructive that those principles, famously encapsulated in the report's paragraph 243, closely reflect the very notions that implicitly informed Recorde's own thinking.[5]

The contributors to the present volume explore aspects of Recorde as a son of Tenby, the content and impact of his publications, and the political and social contexts in which he both lived and worked. The stage is set by Jack Williams, who draws on his extensive research to map out Recorde's biographical details, the nature of his work in the service of the Crown and his interests as an antiquary and linguist. The analysis also highlights Recorde's lack of sufficient political guile to enable him to survive the intrigues of Tudor England, leading to his tragic imprisonment and death.

In the chronological order of their years of publication there follow chapters on each of Recorde's five extant books. John Denniss and Fenny Smith begin by exploring Recorde's *The Ground of Artes* (1543), the first home-grown Arithmetic in English, which establishes a pattern followed by arithmetic texts for almost the next 300 years.

Recorde's only medical publication, *The Vrinal of Physick* (1547), is notable both for its choice of topic – the study of uroscopy – and the use of the vernacular. Margaret Pelling analyses why Recorde, while seeking to establish himself as a medical scholar, chose not to address his treatise in Latin to the select few but sought to reach a wider audience.

Recorde was the first to expound Euclid in English. In *The Pathway to Knowledg* (1551) he had to begin from the very beginning, first to explain what this strange subject of geometry was about and why anyone should trouble to study it, and then to teach its basic rules and constructions. Jacqueline Stedall argues that his work was a notable attempt to convey a significant part of classical learning in English, and was accomplished with considerable literary and poetic style.

The Castle of Knowledge (1556) sets out the elements of astronomy. The subject is portrayed by Recorde as both a humanistic and a practical endeavour, emending classical texts as well as providing instruction for the reader to make an armillary sphere. In addition to analysing Recorde's writing, Stephen Johnston draws on contemporary annotations to consider the book's audience and how they read the text.

In the year before he died, Recorde published his best-known work, *The Whetstone of Witte* (1557), presenting the fundamentals of algebra in English, in which he famously proposed the use of the modern equals sign. Ulrich Reich provides a step-by-step analysis of the way in which the book's arguments are developed and of the sources, mainly by writers in Germany, that influenced its content. Modern readers may be challenged by some of the mathematical intricacies, but they will also gain a strong impression of Recorde's attempts to simplify and to convey the essence of European mathematical thinking to a wider English-speaking audience.

There follow three chapters that explore the historical background and social context of Robert Recorde's work. In the first of these chapters, Nia Powell sets his accomplishments within the background of his birthplace, Wales, and the influence of his upbringing in Tenby, a thriving port with international trading links. Howell Lloyd places Recorde's work within the context of the maritime expansion upon which Tudor England was embarking, a process that led ultimately to the creation of an overseas 'empire' in the modern sense, and analyses both Recorde's successes and failings in supporting this venture. Finally, John Tucker identifies the new role that mathematical ideas and methods were beginning to play in the society and economy of Europe, particularly in relation to the increasingly important role of data and computation, and explores Recorde's engagement with these developments.

Recorde published two other mathematical books as part of his series of texts: *The Gate of Knowledge* and *The Treasure of Knowledge*. No copies of either of these books have survived and there remain only hints in Recorde's extant works as to what they may have comprised.[6] There are also indications that Recorde had further publications in the offing, but his ultimate downfall and early death cut short these ambitions.

We can but marvel at Recorde's accomplishments, while also saddened by the political shortcomings that led to his end. This remarkable 'son of Tenby' bestrode the mathematical world during a tempestuous period in the House of Tudor and emerged, not unscathed, but having made significant contributions to the wide promotion of mathematical ideas, many of which retained their influence and still resonate to this day

At the end of the 'Preface to the Reader' in *The Castle of Knowledge*, in which Robert Recorde extols God's virtues and kindnesses, he introduces a hymn of praise with these words: *'In token therfore of thankfulnes, let vs singe an Hymne vn to that God, praisinge his name, and magnifiynge him foreuer and euer.'*

At a special service in St Mary's Parish Church, Tenby, on Sunday 6 June 2010, marking the 500th anniversary of Recorde's birth, that hymn was sung to more modern words as originally transcribed by Adolf Prag (1906–2004), a teacher of mathematics and Librarian at Westminster School, one of the speakers at the 1958 conference in Carmarthen to commemorate the 400th anniversary of Robert Recorde's death. On Sunday 20 July 1958, the delegates to that conference were transported in two buses to attend a 'Religious Service of Commemoration' at St Mary's Church at which the hymn was sung.

> The world is wrought right wondrously
> Whose parts exceed man's phantasies;
> His maker yet most marvellously
> Surmounteth more all men's devise.
>
> No eye has seen, no ear has heard
> The least sparks of his majesty;
> All thoughts of hearts are fully barred
> To comprehend his deity.
>
> Oh Lord, who may thy power know?
> What mind can reach thee to behold?
> In heaven above, in earth below
> His presence is, for so he would.
>
> His goodness great, so is his power,
> His wisdom equal with them both;
> No want of will, since every hour
> His grace to show he is not loth.

Behold his power in the sky,

His wisdom eachwhere does appear;

His goodness does grace multiply

In heaven, on earth, both far and near.

Notes

1   Frank V. Morley, 'Finis coronat opus', *Scientific Monthly*, 10/3 (1920), 306–8.

2   Geoffrey Howson, *A History of Mathematics Education in England* (Cambridge, 1982), pp. 6–28.

3   Martin Trow, 'The business of learning', *The Times Higher Education Supplement*, 8 October 1993, p. 20.

4   Department of Education and Science, *Mathematics Counts: Report of the Committee of Inquiry into the Teaching of Mathematics in Schools under the Chairmanship of Dr W. H. Cockcroft* (London, 1982).

5   Ibid., paragraph 243 states that 'mathematics teaching at all levels should include opportunities for: exposition by the teacher; discussion between teacher and pupils and between pupils themselves; appropriate practical work; consolidation and practice of fundamental skills and routines; problem solving, including the application of mathematics to everyday situations; and, investigational work.'

6   See, in particular, the verses that appear in the *Castle* (sig. [a.viii]) setting out 'an orderly trade of studye in the Authors woorkes, appertainyng to the mathematicalles'.

ONE

The lives and works of Robert Recorde

JACK WILLIAMS

1. Overview[1]

THE CONFERENCE referred to in the Introduction marked the 450th anniversary of the death of Robert Recorde. It is therefore appropriate to open this first chapter with an account of that event. We have Recorde's undated will, probate for which was first granted on 18 June 1558. A second probate was granted on 6 November 1570, but more of that later. The will was made while Recorde served time as a debtor in the King's Bench Prison[2] at Southwark, south London, unable to pay a fine of £1,000[3] imposed as a result of his having been found guilty of slandering Sir William Herbert, first Earl of Pembroke (1501–70).[4]

The King's Bench Prison was run as a private enterprise with inmates paying for the services provided. Such services included the possibility, under certain circumstances and within specified 'Rules', for inmates to live outside the prison, provided that the services could be paid for. This situation appears to be reflected in Recorde's will, which lists payments of various sorts to prison officials.

The will also lists bequests to members of his family, including one of £20 to 'my widowe mother and stepfather'. Recorde himself was not married and minor goods and chattels were gifted to his cousins and their children. His executors were his brother Richard, and his nephew and namesake, Robert. The cause of his death is not known. However the years 1557–8 were epidemic years in London. In 1558 the country as a whole was smitten with what is believed to have been an outbreak of influenza, which had ravaged the Continent the previous year. Prison would not have been a good place to be during such a season. One of the side effects of such epidemics was that records of births, deaths and burials were not kept, even in parishes more salubrious than Southwark, so it is not surprising that it is not known of what Recorde died or when and where he was buried.

The closest to an obituary for Recorde was written by the physician and author, William Bullein (*c*.1520–1575/6). In the Preface to the second book of his *Bulwarke of Defence*,[5] published in 1562 but compiled earlier while he was in prison, Bullein lists some medical worthies, including Robert Recorde, to whom he devoted most space:

> How well was he seen in tongues, Learned in Artes and in Sciences, natural and moral. A father in Physicke whose learning gave liberty to the ignorant with his *Whetstone of Witte* and *Castle of Knowledge* and finally giving place to eliding nature, died himself in bondage or prison. By which death he was delivered and made free, and yet liveth in the happy land amongst the Laureate learned, his name was Dr. Recorde.

If this list of abilities is augmented by those attributed to him a few years before Recorde's death by the courtier and religious radical, Edward Underhill (1512–*c*.1576),' . . .singularly sene in all of the seven sciences, and a great divine . . .', together they give some feel for the regard in which Recorde was held across the breadth of his many 'lives'.

Evidence relating to the year of Recorde's birth is found in accounts of one of the defining show trials of the reign of Edward VI, that of Stephen Gardiner (*c*.1497–1555), bishop of Winchester, the leading English religious conservative of his time. As a staunch Roman Catholic, Gardiner steadfastly opposed the policy of the Administration to complete the Reformation. To resolve the situation he was asked by the Privy Council to preach a sermon endorsing the religious policy of the regime. He delivered his sermon before king and court on 29 June 1548, but stopped short of compliance with his instructions on a number of issues. He was re-imprisoned, during which time further unsuccessful attempts were made to bring him to heel. Gardiner was then brought to trial at Lambeth on 15 December 1550 and deprived of his bishopric. Depositions relating to the content of the sermon of 1548 were made by members of the Privy Council and their officials, members of the king's court and divines. Twelfth in the list of depositions, sandwiched between those of a high-ranked cleric and a long-established courtier, was one by a 'Dr Robert Recorde, doctor of physicke of the age of 38 years or thereabouts'. The boy from Tenby had travelled a long way geographically, socially and intellectually in his thirty-eight years. If thirty-eight was his age at the time of the sermon then he was born in 1509—10; if at his trial, he was born in 1512.

The Recorde family was deeply embedded in the community of Tenby from well before Robert Recorde's birth, and became even more so after his death. Its genealogy is to be found in Lewys Dwnn's *Heraldic Visitations*.[6] The earliest member of the family noted in the *Visitations* is Roger Record of East Wel in Kent, whose one surviving son Thomas Recorde married twice. Thomas had no children by Joan, daughter and co-heiress of Thomas Ysteven of Tenby Gent. but, by Ros Johns, daughter of Thomas of Machynlleth ap Sion, he had two sons, Richard and Robert. The latter is our subject. Richard married Elizabeth, daughter of William Baenam of Tenby, by whom he had a son and heir named Robert, presumably after his uncle.

Nothing is known of Robert Recorde's youth. How he obtained an education adequate to enter Oxford is a matter for speculation. There is no evidence of the existence of lay schools in the vicinity of Tenby and its church was not collegiate, in the formal sense of that designation. There may have been a modest chantry school, held in the west porch of the church or in the closely associated college.

Robert probably arrived at Oxford about 1525. He was awarded the degree of Bachelor of Arts on 16 February 1531 and elected a Fellow of All Souls in the same year. The annual income from land and properties held by All Souls College in south Wales, Montgomeryshire and Denbighshire allowed it to provide generously funded scholarships and fellowships for Welsh students throughout the period. For example, a holding at St. Clears – only about sixteen miles from Tenby, albeit in Carmarthenshire rather than Pembrokeshire – provided the college with an annual income of about £40 during the sixteenth century.

It seems that Recorde undertook medical studies while at All Souls, but there are no records at Oxford of this activity. The qualification he earned at Oxford for a licence to practise medicine proves that he had, as a minimum, carried out two dissections and effected three cures. There is no firm evidence of when he left Oxford. The probable course of Recorde's subsequent academic career has to be deduced from his records at Cambridge. In 1545 he was awarded the degree of MD by Cambridge on the basis that, by then, he had had twelve years of medical practice after being granted his licence by Oxford to start such activities.

Recorde's subsequent career encompassed a bewildering combination of 'lives' as a mathematician, astronomer and physician, and are discussed elsewhere within these proceedings. This chapter focuses initially on his work in the service of the Crown and its attendant difficulties, then on his interests as an antiquary and linguist and, finally, on what is known about the readers of Recorde's own books.

2. Crown service and its consequences

Robert Recorde and William Herbert, Earl of Pembroke

It was his abilities as an applied scientist that drew Recorde into Crown service and into the unlikely conflict with Sir William Herbert, yet to become Earl of Pembroke. Nowadays Recorde's work for the Crown could be described as that of an iron-founder, a production manager, an accountant and an extraction met-allurgist. All of these activities arise in the context of his dispute with Pembroke and are catalogued in the account of the proceedings that led to Recorde's com-mittal for trial.

Recorde brought the dispute to a head in a letter he wrote on 10 June 1556, during the reign of Queen Mary, to William Ryse, a knight of the Queen's Privy Chamber. Recorde complained bitterly about harassment by Pembroke on issues dating back to the previous reign (that of Edward VI) and of an impending lawsuit. He also included a text in Latin that reads like an extract from a book of prophecies warning of danger to the Crown. He asked to be allowed to answer for his complaints before the Privy Council. Queen Mary passed the letter to certain Privy Councillors and Recorde was called before them only ten days later, when he confirmed that he had indeed written the letter. On 16 October 1556 Pembroke's attorney presented a bill against Recorde alleging that the letter made the Earl appear a traitor, to have injured the Crown and to deserve imprisonment. Damages of £12,000 were claimed.

Recorde replied with a detailed list of allegations, largely involving defrauding of the Crown by Pembroke during the previous reign. Final judge-ment was passed on 10 February 1557 and is merely appended as a brief note to the summary of the case presented to the hearing of October 1556. No detailed record of the trial proceedings has come to light. The judge-ment found that Recorde's letter had damaged the Earl without good reason and awarded damages of £1,000 and £10 costs.[7] It fell far short of what Pembroke had asked. There is however no question of Recorde's financial probity ever being an issue. The alacrity of the royal reaction and the speed of the dispatch of judgement indicated the seriousness with which it was taken. How many members of the Administration other than Pembroke and includ-ing the Crown were involved? Analysis of the accusations levelled by Recorde throws much-needed light on his work for the Crown and on his possibly misguided loyalty to it.

The incident at Pentyrch

The first of Recorde's accusations against Pembroke reads as follows:

From an iron mill at Pentyrch[8] on 3 February 1549 the earl took a barrel of iron worth £10 being Crown property to his own use, and drove off the workers there, to the King's loss of £2000. At Westminster 20 March 1550 the earl persuaded Recorde to give him the profits of the said iron-working which should have been allowed for in the accounts to the profit of the Crown to the value of £200.

The price of iron quoted of about £7 a ton is the rate that held for both wrought and cast iron at the time. Correspondence shows that Pembroke was still pursuing Recorde at the end of 1550 for the balance of the profits he claimed.

There seems to be no reason to doubt Recorde's account of events. If, as the use of the name 'iron mill' suggests, the plant was a bloomery producing wrought iron, then it would have been following the practice at nearby Miskin and it establishes Recorde as being in charge of the earliest known iron-making plant in the lower Taff valley. Assuming a profit of about 50 per cent,[9] corresponding to £3.50 per ton of iron, and using Recorde's figure, a profit of £200 would have arisen from the production of about 60 tons of finished iron, which would have been typical of two years' production of a traditional bloomery. However the capital cost of such a plant would have been far less than the £2,000 that Recorde claimed to have been lost. Such expenditure would have been closer to that of a blast-furnace complex. Whether bloomery or blast furnace, there seems to be no obvious explanation for Pembroke's behaviour. He owned the land if not the mining rights, and had the reversion of such rights. In a previous reign, Henry VIII had been keen to control the use of iron, and here Pembroke had taken Crown property presumably for his own use.

A short digression on monetary matters may help us to understand the nature of the viper's nest Recorde had entered. By the time of Edward VI's accession in 1547 the Crown was almost bankrupt as a result of Henry VIII's military adventures and other extravagances. One means Henry had used to generate revenue was by debasing the currency. The essential details may be summarized as follows.[10] A coin had two values: on the one hand its intrinsic value determined by the weight of precious metal it contained; and on the other, its face value stipulated by whatever the authorities said it was worth. Intrinsic value depends on the weight of the coin and its composition. Time was when a silver penny weighed a pennyweight and was of sterling silver composition, 92.5 per cent silver and 7.5 per cent copper.[11] Towards the start of Henry VIII's reign the weight of a silver penny had been approximately halved. The debasement of the currency was further aggravated by coming off the sterling standard as the percentage of silver was progressively reduced. This provided the Crown with additional revenue, for each time there was a coinage or recoinage it claimed its right to a seigniorage over and above the gain from debasement. Between 1544 and 1551 it is estimated that the total revenue to the Crown from debasement amounted to approximately £1.27m, a sum greater than the total of all Crown taxes and income generated from the sale of Crown lands taken together. The ultimate depth of debasement was reached with the Irish harp pennies whose intrinsic value was less than a seventh of their face value. One practical result of such debasement was a significant rise in inflation – the cost of basic commodities such as bread rose by a factor of three over a period of twenty years. It also provided mint officials with potentially very rich pickings. Corruption was rife at all levels and manifested itself in many ways.

The Bristol mint

Recorde's first appointment as comptroller of the Bristol mint was a direct result of the corrupt behaviour of its under-treasurer Sir William Sharrington, who admitted defrauding the Crown to the tune of about £6,000, to the advantage of both himself and Thomas Seymour, Pembroke's brother-in-law.[12] The appointment was dated 29 January 1549 just a few days before the raid on Pentyrch. Recorde was paid the same annual salary as the new under-treasurer, Sir Thomas Chamberlayne, viz. £133 6s. 8d. The division of responsibilities between them is not clear, but generally the comptroller seems to have had more immediate control of operations. Whatever the split, the outcome was that the mint at Bristol, while operational, turned in an outstanding performance. The quality of product was very high, and when the financial accounts for this period were scrutinized in 1551, they were found to be in surplus to the extent of £218 13s. 2d, an extremely unusual occurrence in mint accounts at the time. The accounts cover the first three-quarters of that year and show a gross profit of some £43,000 and a net profit of a little more than £5,100.

Recorde's charges against Pembroke also accuse him of having played a key role in closing the mint at Bristol, the events being closely associated with the overthrow of Protector Somerset (Edward Seymour, Thomas Seymour's older brother) who had been effectively ruling the country during Edward VI's period as a minor. Pembroke's part in the overthrow has been well explored and opinion on it is still divided. The coup was being organized from London by the Earl of Warwick. The support of Pembroke and Bedford (John Russell, Earl of Bedford) was twice requested by Somerset. They brought their forces as far as Andover by 8 October 1549, from where they responded somewhat coolly to Somerset's request, saying that they were staying put. The following day they retreated a little to Pembroke's manor at Wilton. On 11 October, Warwick (who was made the Duke of Northumberland on that same date) and his followers arrested Somerset at Windsor, where he had fled for safety. Pembroke and Bedford gave their approval to the coup. On 14 October, Somerset was taken to the Tower and Northumberland took over the reins. The evidence deposed by Recorde reveals a sub-plot that exposes Pembroke's attention to raw self-interest. Basically the evidence alleges that, starting on the very day of the coup, Pembroke began to act as if the operation and contents of the Bristol mint were his personal property and lists some four acts by Pembroke to this end.

As a result of Recorde's refusal to co-operate, Pembroke accused him of treason at Westminster on 26 October 1549 and Recorde was forced to 'attend the king' for sixty days from 30 October with a resultant loss to the Crown due to cessation of production at the mint of £10,000. At the house of Sir John Yorke on 6 January 1550, the time and venue for the first meeting of the new Administration under Northumberland, Pembroke tried to bully Recorde into handing over the gold and silver at the Bristol mint, and, when he refused to do so, seized it. In all, Pembroke annexed some £4,000 in coin and bullion. The Bristol mint never reopened. Virtually all of the events related by Recorde in his list of charges against Pembroke can be substantiated and the remainder are supported by circumstantial evidence.

The Dublin mint

Recorde was not out of favour, other than with Pembroke, for too long and soon had a formal involvement with the newly erected Dublin mint. An entry on the Patent Rolls, dated 27 May 1551, shows that Recorde was appointed supervisor[13] of the mint and, by authority of Parliament, was given oversight of all mint officials: the mint's treasurer, comptroller and assayer were charged to heed his counsel in their activities. No one was to coin money without Recorde's consent, any letters patent, commissions or signed bills to the contrary notwithstanding. This was part of a broader commission that involved the operation at Clonmines in County Wexford, which will be discussed later. However, Pembroke still had influence in related matters and the first part of Recorde's case against him relating to Recorde's work in Ireland was that, without consulting or informing him, Pembroke appointed Martin Pirrie to be treasurer of the Dublin mint in May 1552 to coin new money. Following Pirrie's death, he then appointed Oliver Daubney as comptroller to coin further new money between 1 October 1552 and 20 February 1553.

King Edward VI had always taken an interest in matters concerning coinage. By 1552, the finances of the government and consequently those of the king had reached a parlous state. He was seeking to increase his authority and it would seem that for the closing year of Edward's reign the finances of the country were being managed by a small group of people close to him. The Privy Chamber appeared to have taken over financial management from the Privy Council whose members had dispersed to the country. Incoming funds were measured in fractions of a thousand pounds. Revenue from the proposed coinage in Dublin would have been measured in tens of thousands of pounds. Whoever appointed Pirrie, it is clear that the king would have been party to it and have approved, if not initiated, the appointment. It appears to be unlikely that Pembroke would have played any part in this process. The king was aware of the new arrangements for Dublin, as was William Cecil, then one of Edward's two Secretaries of State. Whether Edward initiated the process or whether he was eased in this direction by others is unknown. It is highly unlikely that the Privy Chamber would have felt it necessary to keep Recorde informed. It therefore appears that this set of accusations by Recorde is probably untrue and that Recorde himself was most unlikely to have known the real truth.

Recorde's third accusation, the second relating to the Dublin mint, was that, on 20 June 1553, Pembroke viewed and passed the accounts of Pirrie and Daubney, thereby losing the Crown £40,000. The nature of the indentures under which money was minted at Dublin differed from those extant in England. The king had initially entered into a 'bargain' with Martin Pirrie, then under-treasurer of the Dublin mint, whereby the mint operation was left in Pirrie's control provided he paid the king 13s. 4d (up to a maximum of £24,000) for every pound coined, the indenture expiring in August 1551. Accounts available up to May of that year show that the Crown had received a return of a little over £13,000. There is written evidence that coining beyond August was authorized, but no financial accounts for such an operation exist. Potentially the Crown could have received a further £11,000. The new indenture referred to by Recorde was issued to Pirrie in May 1552 and, coupled with its extension granted to Daubney in December of that year, could have yielded further income to the Crown of £32,400 derived from the coining of 4,500 lb. of fine silver. No financial accounts for this period of operation have been found, so that a total Crown income of between £32,400 and £43,400 remains unaccounted for and is of the same order of magnitude as the £40,000 figure associated by Recorde with the account approved by Pembroke. On this count, Recorde's charge appears to be plausible.

The London mint

Finally Recorde accused Pembroke that, as Supervisor of Mints, he defrauded the king 'in the account of officials' (i.e. by including inaccuracies in the figures) to the sum of £50,000. Not only are accounts for Ireland missing for the period when Recorde had an interest, but those for English gold and silver coining from 1 April 1552 to 24 December 1553 have also not been found. However, a document, probably written in mid-1556, gives an estimate of the production of gold and silver coinage in England from 1542 to 1556. For that part of the reign of Edward VI from Michaelmas 1551 to July 1553 the production of silver and gold is estimated as totalling £145,332 17s. 6d. A separate estimate, made at the start of Elizabeth I's reign in 1558, of the total coinage of gold and silver together struck during the reign of Edward VI was £100,000, leaving an apparent shortfall of coins to the value of £45,000 not in circulation. Were these the depredations that were the basis of Recorde's accusation, and, if Pembroke was not involved, who was responsible, how was it explained at Recorde's trial and by whom? Elizabeth was aware of the discrepancy, attributing it either to hoarding or to exporting. There is no evidence that she took any action against anyone on this count.

The affair at Clonmines

The scene now shifts from the lawcourts to what resembles an episode in the rush for precious metals in the American West. As mentioned above, Recorde's primary function in Ireland was to oversee the silver mining and extraction operation at Clonmines. Interest in such an operation had been initiated towards the end of the reign of Henry VIII. The prospect of finding an indigenous source of silver would have excited any monarch of the time, Henry more than most! On 14 August 1545, armed with a letter of recommendation from the Privy Council to Moris Horner, the Lord Deputy of Ireland, Hans Hardigan and John of Antwerp were dispatched to Ireland to search out mines. The results of their activities were reported in August the following year by Thomas Agard of Ireland, Garret Harman of the City of London and Hans Hardigan. Various 'pieces' both for coin and alum were shown, 'found to be fayre' and were appointed to be kept awaiting the king's pleasure. This pleasure was realized by 25 September 1546, when the Lord Deputy of Ireland and certain members of the Privy Council were authorized to form a Commission for Mines and a prest of £1,000 was ordered for miners to be sent there urgently. The king had however already taken action and had commissioned Joachim Gundelfinger of Augsburg to recruit a surveyor, a smelter and a skilled miner. On 16 March 1546 Gundelfinger reported to Henry that he had recruited the men and had sent them to see Harman, but they had not kept their bargain, to his personal financial loss.

Gundelfinger had served Henry in a variety of ways since 1539 as agent, courier and intelligence gatherer. Harman played a similar role but only from the mid-1540s when, as a London goldsmith with no obvious qualifications for the post, he was appointed overseer of the king's mines. It seems clear that

Gundelfinger and Harman were by that time acquainted with one another, were both foreigners (probably of German origin), were both servants of Henry VIII and both had dealings with members of his inner councils. This is reflected in the account, dated 30 April 1547, of 'Declaration of fees paid by royal warrant out of the Exchequer to foreigners', wherein Harman is listed as being awarded annual payments of £36 10s. for life, Gundelfinger annual payments of £50 for life and Hans Hardigan an optional payment of £40 'at pleasure'. From this time forward their activities became very closely linked to the Clonmines project. However, closeness of interest did not necessarily mean mutual trust and respect. From his letter to Henry of 16 March 1546, Gundelfinger appeared doubtful of Harman's knowledge of mining and extraction metallurgy despite the latter's position as overseer of the king's mines. He wished that Harman had had time to visit the mines at Augsburg to see the methods used there and to take samples of ore to England. He offered his own expertise and sent Harman 'a little book about minerals'. This expertise was to be called upon in 1550.

By the end of 1549 the Clonmines project was again gathering momentum and Gundelfinger received a commission to recruit a team of thirty-seven closely specified miners and smelters. By October 1550 he reported to Cecil that he had arrived in Antwerp with his company of fifty workers and requested financial support and further instructions. Harman arrived in Antwerp on New Year's Day 1551 and writes to Cecil that he has delivered his letter to Chamberlayne at Brussels. In addition Harman happened to have a sample of ore in his bag which he gave to Gundelfinger and the Antwerp Burgomaster. The latter immediately assayed it 'and found it so good that there is no doubt that the King shall receive such honest profit as will cause the Council to regret that it had been so long delayed'. Two days later he writes again to Cecil with the results of assays that had been repeated two or three times. The Burgomaster was very excited about the matter and said that if there was enough ore the whole realm should thank God for it. One hundred ounces of the ore, he reported, should yield more than 8 ounces of silver and half a hundred (i.e. 50 ounces) of good lead. He urged Cecil to show the letter to the Council, who would see that he had always spoken the truth, and 'it were pity that men of no experience should meddle in it as they would lose the one half that God had given to them'. It is not known whether Harman had taken this action himself or whether he had been sent by Cecil without the knowledge of the Council, some members of which were sceptical of Harman's claims. Such sceptics would be justified by the turn of events, even if their judgement was not founded on technical expertise. The sample that Harman had assayed must have contained over 60 per cent galena (lead sulphide, PbS), which level of purity would only have been obtained by a careful selection of samples.

Whatever were Gundelfinger's instructions, they had the required effect, for the miners were in London by the end of May. It is at this juncture that Robert Recorde entered the arena. Recorde's appointment as supervisor of the mint at Dublin also included responsibilities as surveyor of all the king's mines, newly found in Ireland, with commission to rule their affairs including oversight of all matters relating to labour and carriage. Gundelfinger, captain of the foreign miners at Clonmines, and the other officers, workmen and labourers were answerable to him. Recorde also had authority for payment of charges incurred by the mines in the extraction of silver to be coined in the mint at the standard of 4 ounces of silver for every 12 ounces of coinage. Events moved quickly and, by early July 1551, the company was on site at Clonmines and operations had commenced.

Friction between Gundelfinger and Recorde is well documented but the disagreements between them on technical matters seem to have been resolved satisfactorily. Gundelfinger complained about inadequacies of the miners' wages but these rates had been laid down by the Privy Council. There was also friction between Harman and Recorde, probably because the latter seems to have ignored the former, which Recorde probably felt entitled to do as Harman had no official status. Nevertheless, Harman conveyed complaints back to the Council relating to his wishes being ignored.

Meanwhile progress at Clonmines did not augur well. The Council fretted over the delays in production and, when Recorde reported back, it could have been left in little doubt over the financial non-viability of the project. Recorde's report was delayed until February 1552 and it was damning. At that date the king's charges were above £260 a month while his gains were no more than £40, and was losing £220 a month.

The chief cause of the financial debacle was the massive overestimate by Harman of the silver content of the ore to be expected on the basis of the assays at Antwerp. The first trial smeltings produced the equivalent of about 2 ounces of *unrefined* silver from 26 ounces of lead or 100 ounces of ore. Recorde sent a sample of the unrefined silver to Dublin for assay and refining. The purity of the unrefined silver must have been low, for Gundelfinger advised the Council of Ireland that upwards of 4½ ounces of silver might be expected from each hundredweight of pure lead. To obtain 8 ounces of silver, upwards of 4 cwt of ore would have to be treated, which has to be compared with the 100 ounces promised from the same amount of ore by the original assay. Harman persisted in his exaggerated claims in his final letter to the Privy Council, delivered at about the same time as Recorde's report. On 9 May 1552 Recorde presented an account of the total sums of the charges of the mines since 13 April 1551. Two further accounts for the operation were prepared by different auditors, each broadly confirming Recorde's figures.

Recorde prepared final accounts for the period from 20 April 1551 to 11 April 1553. The exact dates of the preparation and presentation of these accounts is not known as the original was said to be lost and we are left with a copy of the account of Thomas Jenison, the king's Auditor General for Ireland at the time, prepared from books submitted by Recorde during the seventh year of Edward VI's reign, and therefore prior to the latter's death in July 1553. Closure of the Clonmines affair had to wait until well into the reign of Elizabeth. Probate of Robert Recorde's will for the second time was granted in November 1570 to his nephew and namesake. The Clonmine accounts were re-audited by a certain L. Jenison, who attested to their accuracy. Successful pursuit of his uncle's debtors by Robert junior for settlement of Robert senior's accounts for the operations at Clonmines was concluded on 12 April 1571. Robert junior was granted the lease for twenty-one years of Crown lands in Cambridgeshire, Sussex, Caernarfonshire and Pembrokeshire having annual rents totalling slightly more than £25 'In consideration that he will relinquish his claim to £1054 19½d owed to his said uncle by Edward VI, as appears by certificate of Thomas Jenyson, Auditor of Ireland, remaining with the Queen's Remembrancer at Westminster'. Robert senior had thereby secured the well-being of his family in Tenby.

During this period between 1549 up to his death, while Recorde was engaged in these worldly activities, he was also, remarkably, preparing edition B of his *The Ground of Artes* and his three later texts. The failure of the Crown to settle his accounts must have left him very short of money at a time when, following the accession of Mary, his overt Protestantism hardly commended him to the authorities.

3.   Antiquary and linguist

Recorde probably began to indulge in antiquarian activities during his time at Cambridge. One of the least desirable outcomes of the Reformation was the vast loss of manuscripts from college and monastic libraries. One of the reactions to this vandalism was the subsequent growth of antiquarian activities. It started with John Leland (*c.*1506–52) and continued with John Bale (1495–1563). Bale's catalogue of British writers includes evidence that allows us to classify Robert Recorde as an antiquarian forerunner of Archbishop Matthew Parker (1504–75) and his chaplain and amanuensis, John Jocelyn. Books classified as 'ex museo Robertus Recorde' form one of the three largest collections listed by Bale. The titles cover a wide range of interests: religion, medicine, alchemy, prophecy, heraldry, geography, history, arithmetic, astronomy and law. Excluding Recorde's own books, complete or projected, the list includes upwards of ninety items all of which, with one exception, were in the form of manuscripts.

The largest collection of titles in Recorde's collection concerned astronomy and in content provides a roll-call of English authors writing on this subject. The oldest of these documents, it is asserted, is the *Theorica planetarum* ascribed by Bale to Roger of Hereford. The astronomical tables and almanacs compiled by the Merton school in the fourteenth century are represented by the works of John Mauduith, Simon Bredon, John Ashenden, John Somers and William Rede. Other fourteenth-century astronomical texts include those by John Walters and John of Northampton. The limited activity by English astronomers in the following century are represented by the tables of John Holbrook at Cambridge and, much later, by the eclipse calculations of Lewis of Caerleon, described by Bale as 'fragments plus a book'. Two members of the Merton school, John Kyllingworth and Simon Bredon, also provided texts on arithmetic.[14] Bale also lists three books by Richard of Wallingford in Recorde's library, viz. the descriptions of the instruments called Albion and Rectangulam and his *Extrafrenon*, a judicial astronomy.

In a letter dated 30 July 1560 to the newly appointed Archbishop Matthew Parker, Bale names individuals who held collections of 'bookes of antiquite not printed'. The list includes Sir John Cheke, Lord Paget, Sir John Mason, Lord Arundel, John Twyne, Robert Talbot and the executors of Robert Recorde. Whatever happened to his manuscripts subsequently, Recorde had provided a safe haven for them during a perilous period of history.

Recorde has provided us with over 1,000 pages of printed text but only one set of examples exists in what may have been his own handwriting, and that in the form of annotations in Anglo-Saxon. On the flyleaf of a volume of MS 138 of the *Chronica varia*[15] now at Corpus Christi College, Cambridge is found the inscription '*Robertus Recorde erat qui notauit hunc librum Characteribus Saxonicis*.'[16] Studies have concluded that Recorde had taken some of his material from Cotton Tiberius B.1,[17] one of the annals of the Anglo-Saxon Chronicle, but that the Genealogical Preface which Recorde had copied into MS 138 had probably originated in MS Kk 3.18, now at Cambridge University Library. Both manuscripts had been owned by Robert Talbot (*c*.1505–58), a fellow antiquarian. Leland was a friend of Talbot and had borrowed Cotton Tiberius B.1 from him. Talbot had entered New College, Oxford in 1521 and supplicated for his MA in 1529. He was therefore a contemporary of Recorde at Oxford. Recorde may also have used Henry of Huntingdon's *Historia Anglorum*, Florence of Worcester's *Chronicon* and his own copy of Roger of Hovedon's *Chronicon* as sources written in Latin for some of his annotations to MS 138.

Cotton Tiberius B.1 is also likely to have been the source of historical material described by Recorde in the Dedicatory Epistle to *The Whetstone of Witte*. Speaking to the dedicatees, the Muscovy Company, about his projected book on navigation he says (*Whetstone*, sigs a.iii–[a.iii]ᵛ):

> Wherein I will not forgett specially to touche, bothe the olde attempte for the Northlie Nauigations, and the later good adventure, with the fortunate successe in discoueryng that voiage, which noe men before you durste attempte, sith the tyme of kyng Alurede his reigne. I meane by the space of. 700. yere. Nother euer any before that tyme, had passed that voiage, excepte onely Ohthere, that dwelte in Halgolande: whoe reported that iorney to the noble kyng Alurede: As it doeth yet remaine in aunciente recorde of the olde Saxon tongue.

This extract has the distinction of referring in print, for the first time, to an Anglo-Saxon source.

John Kyngston, who printed *The Whetstone of Witte*, also printed the fourth edition of Fabian's *Chronicle* (1559). This edition of that work describes itself as 'newley perused' and it seems probable that the peruser was Recorde, who was responsible for a substantial annotation relating to the genealogy of Celtic kings and for an even greater contribution, which the peruser references as being derived from Geoffrey of Monmouth, of whose works Recorde had a copy.

Translation from Latin, Greek, and Arabic (to a very limited extent) as well as from other vernacular languages into English of topics requiring their own specialist vocabulary, must have posed particular difficulties to Recorde. By his own telling, he had to retire hurt on at least one occasion. One particular comment bears repeating, and was made in the context of Recorde's use of Latin words for terms describing various types of proportionalities (*Whetstone*, sig. [B.iv]ᵛ):

> *Scholar.* Why doe you not name theim all by Englishe names?
> *Master.* Bicause there are no soche names in the Englishe tongue. And if I shoulde giue theim newe names, many would make a quarrelle against me, for obscuryng the olde Arte with newe names: as some in other cases all redy haue doen.

Here Recorde must have been referring to the failure of acceptance of his proposed geometrical terms. However a need for original English nomenclature surfaced later with the subject matter of the *The Whetstone of Witte* particularly with respect to the classification of numbers. Here are found for the first time a whole array of designations and definitions, including: numbers abstract, contract or denominate, relative, figural, simple or uncompound, compound, perfect,

superfluous or abundant, diminutive or defective, commensurable, incommensurable, linear, superficial, diametral, square, triple, quadruple, multiplex, manifold
and absurd. To develop his algebra Recorde adopted *cossike* terms from German
and Latin texts but he also introduced other new terms, including equation, binomial and bimedial. Some of these terms have since fallen by the wayside, but
many are still in use today and, while Recorde is more popularly known for his '='
sign, he should also be credited with their introduction as well. To this list should
also be added the use of the word 'sign'.

4. Recorde's readers

Recorde's works had a wide readership, including some eminent and interesting
names, and some specific observations can be made on the basis of the evidence
available. For example, it is reasonable to expect that editors of any one of his
books would be conversant with his other volumes. As the first editor of *The
Ground of Artes*, John Dee must also have read *The Whetstone of Witte*, for he uses the =
sign several times in his much-lauded *Mathematicall Praeface* to Billingsley's *Euclid*.[18]
It is, however, difficult to understand the absence of Recorde's texts, other than
a second edition of *The Vrinal of Physick*, from Dee's extensive library. Although
Dee was part of Pembroke's household in 1552 and of Northumberland's the
following year, he showed little overt regard for Recorde and may even have suf-
fered from professional jealousy. John Mellis, the second editor of *The Ground of
Artes*, was unstinting in his praise for the book, attributing to it the mathematical
enlightenment of thousands.

One person who profited was Richard Norwood (*c.*1590?–1675), mathematician and surveyor, who wrote: 'It was while I was in this employment [on a
sailboat] that having by me Record's Arithmetic given me a little before by my
father I went through it in numbers and fractions in some three weeks' time
whilst we lay at Yarmouth.' Norwood became an eminent author in his own right,
was the first man to achieve an accurate measure of the nautical mile and had a
significant role in the history of Bermuda. Sir Martin Frobisher (1535?–94),
the English privateer, navigator, explorer, and naval officer, took a copy of *The
Castle of Knowledge* with him to America – an early example of distance learning!
The reader whose comments Recorde would probably have valued most was John
Aubrey (1626–97), antiquary and writer, who devoted several pages to Recorde
in his compilation *Brief Lives*. His copies of three of Recorde's books are to be
found in the library of Worcester College, Oxford, *The Ground of Artes* being annotated. Aubrey was a competent mathematician and a Fellow of the Royal Society,
so his very favourable comments carried some weight.

5. Postscript

The absence of personal archival material makes it difficult to get a real feel for what sort of a man Robert Recorde was. However the man does peep out occasionally from behind the curtains of his formal writings and there are a number of asides or one-liners that give him a human face and display insights that continue to have relevance. A steady favourite of mine remains (*Ground*, 1558, sigs. [a.vii] ᵛ–[a.viii]): 'Oh in how miserable case is that realme where the ministers and interpreters of the lawes, are destitute of all good sciences, whiche [a]re the keyes of the lawes? Howe can they either make good lawes, or mayntayne them, that lacke that true knowledg, whereby to iudge them?' This may be coupled with the observation (*Castle*, p. 127 [129]), resonant of Alexander Pope's 'a little learning is a dangerous thing': '. . . more often haue I seen it by experience, that better it is for men to want all arte of reasoninge cleane, then to haue suche confidence in a meane knowledg therof, that may occasion them to deceaue them selfe, and to seduce other.'

Neither was he short of skills as a versifier. Justifying the study of astronomy, Recorde writes (*Castle*, sig. a.iiij):

> If reasons reache transcende the Skye,
> Why shoulde it then to earthe be bounde?
> The witte is wronged and leadde awrye.
> If mynd be maried to the grounde.

On teaching methods he comments (*Whetstone*, sig. [G.iiii]) that 'examples are the lighte of teachyng' and 'how greate a helpe the sighte of the eye doth minister to the righte and speedye vnderstandyng of that, whiche the eare doth heare' (*Castle*, p. 66). My current favourite is his response (*Castle*, p. 141 [241]) to a weak argument put forward by Sacrobosco: 'That reason sauoreth more of the determinations theological, then of the demonstrations mathematical.' His combination of wit and insight guarantees a good read.

Robert Recorde was a remarkable polymath, endowed with exceptional communication skills. Sadly, these skills did not extend to the political sphere where, as we have seen, his lack of guile and his meticulous attention to accuracy, regardless of the consequences, ultimately led to his demise.

Notes

1   For a more detailed and fully referenced account by the author of Robert Recorde's life and work, see Jack Williams, *Robert Recorde: Tudor Polymath, Expositor and Practitioner of Computation* (London, 2011). This chapter is a condensed version of parts of that fuller account.

2   The King's Bench Prison took its name from the King's Bench court of law in which cases of defamation, bankruptcy and other misdemeanours were heard. Accordingly, the prison was often used as a debtor's prison.

3   A factor of about 300 has to be used to convert money values of the period to those of today.

4   William Herbert was knighted in 1544 by Henry VIII, made Knight of the Garter in 1549 by Edward VI, created Baron Herbert of Cardiff on 10 October 1551 and 1st Earl of Pembroke the following day. He is referred to as Pembroke throughout this chapter.

5   *The Bulwarke of Defence* comprises five books, the second of which is entitled *A Dialogue betwene Sorenes and Chyrurgi on Apostumacions and Wounds.*

6   Lewys Dwnn (c.1550–c.1616), of Betws Cedewain, Montgomeryshire. See his *Heraldic Visitations of Wales and Part of the Marches*, ed. Samuel Rush Meyrick (Llandovery, 1846).

7   The documentation was found in a King's Bench roll in the 1960s by W. Gwyn Thomas (1928–94), to whom credit for the discovery of the archive is acknowledged.

8   The village of Pentyrch is situated five miles to the north-west of Cardiff on a ridge to the south of Mynydd y Garth.

9   David Crossley's paper on profits from contemporary iron making at Robertsbridge in the Weald of Kent provides the basis for this assumption. See D. W. Crossley, 'The management of a sixteenth-century ironworks', *Economic History Review*, 19/2 (1966), 273–88.

10   An authoritative and detailed description is given by C. E. Challis, *The Tudor Coinage* (Manchester, 1978).

11   This new standard for English coinage was established by Henry II in the twelfth century.

12   Pembroke's first wife, Anne Parr, was a sister to Catherine Parr, the last of the six wives of Henry VIII, who married Thomas Seymour in 1547, shortly after Henry's death.

13   The terms 'surveyor of the mint' and 'supervisor of the mint' are used synonymously according to the terms of individual indentures.

14   These items belonged originally to Lewis of Caerleon and are now to be found in Cambridge University Library, MS Ee. 3. 61.

15   The inscription occurs at the end of a Table made for Archbishop Parker. The Parker Library at Corpus Christi College, Cambridge, houses a collection of over 600 manuscripts, particularly medieval texts, the core of which was bequeathed to the College in 1574 by the Archbishop.

16   Loosely translated as 'Robert Recorde has annotated this book in/from Saxon characters.'

17   Currently kept at the British Library.

18   The first translation of Euclid's *Elements* into English was made by Henry Billingsley, *The Elements of Geometrie of the Most Auncient Philosopher Euclide* (London, 1570).

TWO

Robert Recorde and his remarkable Arithmetic

JOHN DENNISS AND FENNY SMITH

IT IS NOT JUST for the alliteration that we have inserted the word 'remarkable' into the title of this chapter; we believe it is truly deserved. As far as we know, it was the first home-grown Arithmetic in English; it was preceded by Cuthbert Tunstall's text of 1522 in Latin,[1] and by the anonymously written *An Introduction for to Lerne to Rekyn with the Pen and with Counters*, first published in 1536/7, which, although in English, was largely a translation from Dutch and French texts of the same title.

Recorde gave highly imaginative titles to all his works, grabbing the potential reader's attention and, in the case of *The Ground of Artes*, implying that arithmetic was the foundation for many other studies: it was, indeed, one way by which to distinguish humans from other animals. As stated in his Preface (p. 6):[2] 'for who so setteth small pryce by the wittie deuyse [devise] and knowledge of numbrynge, he lyttell consydereth it to be the chyefe poynt (in maner) wherby men dyffer from all brute beastes . . .'

Recorde devotes the first ten pages of his book to explaining the need for arithmetic, including this passage (sigs A.ii–A.iiᵛ):

> *M.* If nombre were so vyle a thynge as thou dyddest esteme it, then nede it not to be vsed so moch in mens communycation. Exclude nombre and answere me to this question: Howe many yeares olde arte thou?
> *S.* Mum.
> *Master.* Howe many dayes in a weke? how many wekes in a yere? what landes hath thy father? howe many men doth he kepe? how longe is it syth you came from hym to me?

Sco Mum.

M. So that yf nombre wante, you answere all by mummes: Howe many myle to London?

Sco. A poke full of plumbes.

An Arithmetic with humour in it is something of a rarity! The passage demonstrates the second remarkable feature of the book, that it was written in dialogue form, which, although having a long history dating back to Plato, was rarely used in practice. Recorde breathed new life into it and his choice was entirely appropriate, as he explains in his Preface (p. 8): 'I haue wrytten in the fourme of a dyaloge, bycause I iudge that to be the easyest waye of enstruction, when the scholer may aske euery doubte orderly, and the mayster may answere to his question playnly.' At a time when access to education was severely limited, this was a masterstroke, making the learning of arithmetic accessible to anyone able to read, whether they had a tutor or not.

Recorde wrote *The Ground of Artes* in two parts, the first dealing with whole numbers, probably published originally in 1543 (we have not been able to trace any earlier copy), and the second part dealing with fractions no later than 1552.[3] Recorde died in 1558, but these two parts, in the original black-letter type typical of the period, remained virtually intact through almost all the subsequent editions, such was the respect in which they were held by subsequent editors. Further additions were made, particularly a third part in 1582, but in Roman type. And it stayed the course. There were some forty-five printings over a period of more than 150 years, the last being in 1699, and this at a time when books were expensive in relation to wages and the cost of living. So what was it that gave it such appeal?

Firstly, it was almost all exposition with very few exercises – the latter came into prominence in the eighteenth and the nineteenth centuries. As a comparison we might look at another best-seller, namely Walkingame's *Tutor's Assistant*, first published in 1751, 70 per cent of which was devoted to exercises as opposed to about 1 per cent by Recorde. The inclusion of exercises was one of the most significant subsequent developments in arithmetical textbooks, and paralleled the development of education, as teachers began to make significant use of textbooks as sources of exercises. Recorde was writing mainly for those who were teaching themselves, scholars who would have no one to check their answers to the exercises.

So what was actually in the book? The four basic operations, of course, formed the framework: addition, subtraction, multiplication and division. Subtraction provides a good example of the clarity of Recorde's explanations. There are two main methods of performing subtraction calculations in vertical format where the lower digit is larger than the upper as in, for example, subtracting 17 from 52, where the unit digit 7 in the lower number is larger than the unit digit 2 in the upper number:

(a) Decomposition

$$52 - 17 \rightarrow (40 + 12) - (10 + 7)$$
$$\rightarrow (40 - 10) + (12 - 7)$$

(b) Equal Additions

$$52 - 17 \rightarrow (50 + 12) - (20 + 7)$$
$$\rightarrow (50 - 20) + (12 - 7)$$

Recorde chooses the second method (sig. E[.viii]):

> If any figure of the nether summe be greater then the figure of the summe that is ouer hym, so that it can not be taken out of the figure ouer hym, then must you put 10 to the ouer figure, and then consyder how moch it is: and out of that hole summe withdrawe the nether figure, and write the reste vnder them.

After the Scholar's response, he goes on (sigs [E.viii]–[E.viii]^v): 'So haue you done well, but now must you marke another thyng also: that when so euer you do put 10 to any figure of the ouer nomber, you muste adde one styll to the figure or place that foloweth nexte in the nether lyne . . .' This is a very nice description of the method of equal additions, one which authors of some of the subsequent Arithmetics would have done well to copy.

Later in the same example, the Scholar repeats the method by saying 'Then shall I saye bycause I borowyd 10 to the ouer 5, I must put 1 in the nexte place beneth' (sigs F[.i]–F[.i]^v), in which Recorde invokes the word 'borrow', a term that already had some currency[4] and which, despite its semantic ambiguity, has continued to be used at least in the informal vocabulary of subtraction to the present day.

Recorde's sensitivity to providing teaching aids that are accessible to his readers is well illustrated by his discussion of the multiplication of single-digit numbers (sigs G.v^v–[G.vii]). He firstly asserts that there is no need to teach a rule for multiplying single-digit numbers that are each below six because 'they are so easy that euery chyld can do it.' The multiplication of larger numbers (between 5 and 10) is a different matter and Recorde explains his method as it applies to finding 8 × 7, starting by placing the numbers to be multiplied one under the other, thus:

$$8$$
$$7$$

'Then do I loke how moche 8 doth dyffer from 10, and I fynde it to be 2, that 2 do I write at the ryghte hande of 8.' Similarly 7 is subtracted from 10 to give 3. 'Then do I drawe a lyne vnder them':

8	2
7	3

The two differences (2 and 3) are now multiplied to give 6, placed under the line in the 'units' column. Finally, 'must I take the one of the differences (which I wyl for all is lyke) from the other digette, not from his owne.' Thus we find the difference 7 − 2 (or 8 − 3) to give 5, which is placed under the line in the 'tens' column:

8	2
7	3
5	6

Thus 8 × 7 is shown to be 56.

This teaching aid is mentioned by a number of fifteenth- and sixteenth-century British and north European authors, including Peurbach (1450), who speaks of it as an ancient rule, Huswirt (1501), Riese (1522), Rudolff (1526) and Stifel (1544). It was also commonly used as a method of finger reckoning, surviving in parts of eastern Europe well into the twentieth century. In this

Figure 1.   Recorde's table of basic multiplication facts

method, for example, to multiply 7 by 8, raise 3 fingers on one hand and 2 on the other (the respective complements in 10). Multiply the raised fingers for the units and add the closed fingers for the tens, to obtain 56.[5]

In order to assist the Scholar further, Recorde sets out 'A table to multiply all dygettes by' (sig. [G.vii]^v). This table (see Figure 1) sets out the minimum collection of products that need to be known and, as a concise statement, was never bettered. Recorde provides precise instructions for the Scholar to use the table:

> In whiche figure when you wolde knowe any multiplication of digettes, seke your fyrste or laste digette, in the blacke squares, and from it go ryght forth towarde the ryghte hande, tyll you come vnder the figure of your seconde diget whiche is in the hyghest rowe, and then the nomber that is in the metynge of theys bothe squares, is the multiplycation that amounteth of them.

After a section on *Reduction* (the process of reducing large units to smaller denominations, for example, £ s. d to farthings) and the reverse process, Recorde provides a chapter on *Progressions*, exploring both arithmetic and geometric progressions, including their sums. Odd though it may seem to us to treat such an advanced subject this early in the book, there is a long tradition that regards *Progression* as one of the fundamental operations. For example Sacrobosco (*c.*1250) lists nine basic operations (numeration, addition, subtraction, duplation, mediation, multiplication, division, progression and root extraction) in his *Algorismus*, one of the first works to introduce the Hindu–Arabic numerals into Europe, although Pacioli (1494)[6] reduced the number (with the omission of duplation and mediation) to seven, and Gemma Frisius (1540) to our traditional four. By Recorde's time, Pacioli's *Summa* had been widely dispersed throughout Europe and studied by many sixteenth-century mathematicians,[7] and the fact that Recorde follows Pacioli's arrangement here and elsewhere shows that he too was influenced by the *Summa*.

Recorde provides two worked examples on this subject, both based on geometric progressions. The first (sigs O[.i]– O.ii) is the well-known horseshoes problem (an *ob* is a halfpenny):

> Yf I solde vnto you a horse hauyng 4 shoes, and in euery shoe 6 nayles, with this condition; that you shall pay for the fyrste nayle 1 ob for the second 2 ob. for the thyrde 4, and so forth, doblyng vnto the ende of all the nayles. Nowe I aske you howe moche wolde the hole pryce of the horse come vnto.

Eventually the Scholar arrives at the staggering price of £34,952 10s. 7d and an ob! The ludicrousness of this answer is well recognized:

M. That is well done : but I thynke you wil bye no horse of the pryce.

S. No syr, yf I be wyse.

More wry humour.

The second example (sigs O.ii– O.iii^v) also leads to an unrealistic answer, although Recorde provides no commentary on this occasion. The example concerns the task of laying bricks to make a series of twelve walls. The task is to determine how many brick loads are delivered to the bricklayer, although what constitutes a 'lode of brycke' is never made clear. The Scholar is told that each successive wall uses up two-thirds of however many brick loads remain and that only one brick load is left unused at the end of the exercise. His initial response is that it is not possible to do the calculation: 'Why Syr? it is vnpossyble for me to tell.' The Master proceeds to instruct the Scholar, based on an initial observation that using up two-thirds of the brick loads at each step necessarily leaves one-third unused. He then works backwards, noting that the bricklayer must have had three brick loads at his disposal in order to build the twelfth and final wall, using up two of them and leaving the third as the unused remainder. Using a similar process, he reasons that there were nine brick loads before the eleventh wall was built, 27 brick loads before the tenth wall was built, 81 brick loads before the ninth wall was built, and so on, finally deducing that there were 531,441 brick loads before the first wall was built and illustrating the answer in tabular form. The Scholar is greatly impressed: 'Lo, nowe it appeareth easy ynoughe. Now surely I se that Arithmetike is a ryghte excellent arte.' The Master builds on this enthusiasm by promising more to come, to which the Scholar responds, 'Then I beseche you syr, sease not instruct me farther in this wonderfull connynge.'

The next section (sigs [O.iii]^v–Q.ii) is the core of any Arithmetic of the time: the famous, or infamous, rule of proportions or *Golden* rule. This deals with simple proportion. For example, if 6 articles cost 10 pence what would 9 articles cost at the same price? The Rule for dealing with this type of problem was usually stated as 'Multiply the second and third numbers together and divide by the first'. So in this example we would have 10 × 9 is 90, divide by six giving the answer 15 pence.

Using Recorde's diagrammatic representation, this would appear as:

Recorde describes the Golden rule as being one 'whose vse is by 3 nombers knowen, to fynde out another vnknowen, which you desyre to knowe'. His own introductory example (sigs O.iii^v–O.iiii) concerns the payment of rent: 'If you paye for your borde for 3 moneths 16 s. how moch shall you paye for 8 monethes.' Referring to his diagram, his rule is expressed as 'I must multiply the lowermost on the lefte syde by that on the ryghte syde, and the summe that amounteth I must diuide by the hyest on the lyfte syde'. Thus, for our simpler example above, we perform the calculation 9 × 10 then divide by 6 to give 15 as the answer to complete the diagram. The rule serves him well to solve a number of problems and later editions of the book are amended by editors to elaborate further on the varieties of examples for which what also became known as the *Rule of Three* or the *Rule of Three Direct* can be applied.

The trouble with using rules is that you have to know the right one to use in any particular case. If the problem is worded differently or the situation is varied, a different rule may be needed. Recorde provides a word of caution that the Golden rule does not apply in all cases and that a second rule also needs to be learned for 'there is another order quyte contrarye to this that you haue learned' (sig. [O.vi]). This new order, or rule, he terms the Backer rule, subsequently in later editions to be referred to as the *Rule of Proportion Backward*, or *Reverse*. To illustrate his argument, Recorde presents an example (sigs [O.vi]–[O.vi]^v) concerning measurement of cloth: 'Yf I have bought 30 yardes of cloth, of two yardes bredthe, and wolde bye canwas of 3 yardes brode to lyne it with all, how many yardes shulde I nede?' The hypothetical pupil's immediate reaction is to challenge the reality of the question: 'Why? there is none so brode.' To which the Master rather testily replies: 'I do not care for that, I do putte this example onely for your easy vnderstandyng : for yf I shulde put the example in other measures, it wold be harder to vnderstand . . .'

Here, there are still three quantities involved but the *Rule of Three Direct* does not apply; instead, we need the *Rule of Proportion Backward*, or *Reverse*, set out by Recorde as 'you shall multipye now the fyrste nombre by the seconde, and that aryseth therof, you shall diuide by the thyrde' (sig. [O.vi]^v) leading to the calculation: 2 × 30 then divided by 3 to give 20, as the answer, set out by Recorde in tabular form as:

Recorde also considers problems that involve five numbers, such as this example (sig. P.ii), whose form may be familiar to some readers from their own schooldays: 'If 6 mowyers do mowe 45 acres in 5 dayes, how many mowyers wyll mowe 300 acres in 6 dayes?' To solve this problem we need to apply the technique twice to form what Recorde calls a *Double Rule*. In this particular case we apply the Golden rule followed by the Backer rule – but different problems require different combinations of the rules. Recorde is well aware of the danger of over-reliance on rules as is shown in a significant exchange between Master and Scholar earlier in the volume (sig. [F.vii]ᵛ):

> *S.* Syr I thanke you: but I thynke I myghte the better do it, yf you dyd shewe me the workynge of it.
>
> *M.* Yea but you muste proue your selfe to doe some thynges that you were neuer taught, or else you shall not be able to do any more then you were taught, that were rather to learne by rote (as they call it) then by reason.

It is also significant that Recorde appears to have chosen his numbers carefully in the example involving mowers because the answer is not a whole number, thus allowing him to introduce the notion of part of a number, as he explains (sig. P.iiᵛ): 'But suche broken nomber, called fractions, you shall here after more better perceaue, when I shall holy enstructe you of them.' Recorde fulfils this promise by including fractions in his revised edition of the *Ground* (see below).

There are other chapter headings in *The Ground of Artes* that are unfamiliar to us. *Fellowship*, for example, which, in its simplest form, concerns merchants who invest different sums of money in an enterprise and expect to get the appropriate share of the profits.[8] That's straightforward, of course, but it gets a little more complicated in the *Rule of Fellowship with Time*, when the merchants invest money for different lengths of time, such as in this problem involving four merchants (sig. P.v):

> iiii. marchauntes made a commen stock, which at the yeares ende was encreased to 35145 li. [pounds] Now to knowe what shalbe eche mannes porcyon of gaynes, you muste knowe eche mans stocke and tyme of continuaunce. The fyrste man of these .iiii. layde in 669 li. which he dyd take from the stocke agayne at the ende of 10 monethes. The second man layd in 810 li. for 8 monethes. The thyrd layde in 900 li. for 7 monethes. And the fourth layde in 1040 li. for 12 monethes.

At this point Recorde begins what he calls *The seconde dialoge*, a quite different way of doing arithmetic, with counters and a counting-board (sigs Q.iiᵛ– T.ii). This has ancient origins, was still of importance in the sixteenth and seventeenth centuries and, indeed, persisted into the twentieth century. Apart from in the nursery, the use of the bead abacus in modern times has been predominantly confined to Russia and the Far East. In northern Europe, the preference has been for counters in conjunction with a counting board, and special counters continued to be minted for this purpose until well into the eighteenth century.[9]

Recorde gives examples of each of the four basic operations using a counting board. For instance, the diagram in Figure 2 represents the first of eight stages in the calculation of 1542 × 365, a problem clearly linked to the year in which the book was written, if not published.

Figure 2. Counters on a counting board showing the first stage
in the calculation of 1542 × 365

A counter placed on the starred line represents the number 1,000; a counter on the line beneath represents 100, and so on. A counter placed midway between two lines is equivalent to five counters on the line below. The number to the left of the vertical line therefore represents 300 + 50 + 10 + 5 = 365; the number to the right represents 1000 + 500 + 40 + 2 = 1542. This method was profligate in the use of space – eight diagrams and seven pages to do just one multiplication sum. What was really required, of course, was a hands-on demonstration, but this was the next best thing. The Master draws the example to a close by concluding that '1542 (which is the number of yeares syth Ch[r]ystes incarnation) beyng multyplyed by 365 (which is the number of dayes in one yeare) dothe amounte vnto 562830 . . .'

The section on counters, which disappeared from *The Ground of Artes* in editions printed after about 1673, was followed by a short section on *The arte of nombrynge by the hande* (sigs [T.iii]–[T.vi]). This is usually described as 'finger notation', shown in Figure 3, and is a system for displaying numbers on the digits of both hands (hence the origin of the word digit). According to Recorde the origin of this system goes back at least 2,000 years, and certainly the Venerable Bede described it in 707. What was the point of this system? It seems to be that it was language- and symbol-free; anybody from anywhere could tell how many sheep they wanted to buy without uttering a word. It was still in the edition of 1582, but was discarded soon after.

The original book ended at the end of this section, after some 317 pages. By the 1552 edition of the *Ground* Recorde had added *The Second Part of Arithmeticke*

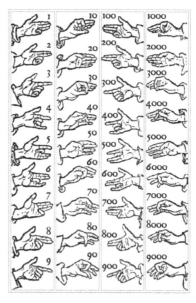

Figure 3. An ancient system for displaying numbers using both hands

tovchyng fractions, brefely sette foorthe.[10] Not that briefly, however, as he added fifty-three pages in which he went through addition, subtraction, multiplication, division, reduction and the Rule of Three all over again with fractions. Besides this additional work on fractions, Recorde also added a further fifty-four pages including sections on *Alligation* (see below), which is concerned with problems such as the mixing of wines of different values in quantities that gave a product at a predetermined price, and the *Rule of Falsehood* (1552, sigs [X.vi]–a.i), of which this is an example:

> A mason was bound to build a wall in 40 daies, and it was couenanted so with him, that euery day that he wrought he shoulde haue for his wages 2s 1d, and euery day that he wroughte not, he shoulde bee amerced 2 shyllynges syxe pens, so that when the wall was made, and the recknyng taken of the dayes that he wrought, and of the other that he wrought not, the mason had clearely but 5s.5d for his worke. Now do I demand how many dais dyd he work of those 40, and how many did he not woorke?

This is a problem crying out for a bit of simple algebra, but Recorde could not, any more than his predecessors, assume any such knowledge on the part of the reader even though he himself had a good grasp of it.

34

The *Rule of False* is a traditional technique, known to the ancient Egyptians, which reached Europe from India and Arabia via the works of Leonardo (Fibonacci) of Pisa (1202).[11] Its popularity rested on the fact that it was purely arithmetical: one or two guesses were made for the required value, and the resulting errors were adjusted according to a given rule, based upon proportion, to give the correct answer. There were two rules, the 'rule of (simple) false' for when there was one unknown, and 'double false' for when there were more than one. Accordingly, Recorde gives an arithmetical procedure, equivalent to the following, for arriving at the answer:

(a) As a first step take some number of days 'at aduenture' that he worked, say, 28: then he 'played' for the remaining 12 days. Now calculate his wages, which would have been 340 pence. This is far too much, being 275 pence more than his actual wages.

(b) For the second step choose another combination of numbers, for example that he worked for 24 days and played for 16. This would give him 120 pence. This is still more than the 65 pence he actually earned, this time by 55 pence.

(c) Can we now use these two initial guesses to calculate the correct answer? To do this, Recorde uses an ingenious algorithm. To set up the algorithm he forms a cross ('commonly caled saint Andrewes crosse') at the points of which he writes, in a specified order, both the numbers of days the mason 'played' and the corresponding resulting errors, as follows:

(d) The calculation now proceeds by cross-multiplying and finding the difference between the answers. In this example $275 \times 16 - 12 \times 55 = 3740$. The difference between the two errors is now calculated, being, in this case, $275 - 55 = 220$. Finally the first of these answers is divided by the second to give the correct number of days played. Thus 3740 divided by 220 is 17, from which we 'deduce' that the mason played for 17 days and that he therefore worked for 23 days.

Recorde sets out the whole method using particular examples over a total of forty pages, including consideration of special cases. This is a sure-fire, if a rather lengthy, method. In order to maintain the interest of his readers and, possibly, as an aide-memoire, Recorde adds a sixteen-line 'obscure ryddle' that ends with the couplet:

> Gesse at this worke as happe doth leade.
> By chaunce to truthe you may procede.

The Rule of Alligation (1552, sigs [W.vi]–[X.vi]), another traditional technique to do with valuing mixtures of constituents of different values, such as fruit, wine or even metals for coinage, brought Recorde a new problem – how to introduce his Scholar to the idea of indeterminate equations without mentioning them! His examples, set out over sixteen pages, deal with 'diuers parcels of sundry quantities', such as the following, which involves six variables connected by only two equations:

M. ... A marchant beynge mynded to make a bargayne for spices in a myxt masse, that is to saye, of cloues, nutmigges, saffron, pepper, ginger, and almondes, the cloues beeyng at 6 s a pounde, the nutmigges at 8 s, saffron at 10.s, pepper at 3 s. gynger at 2.s. and almondes at 1.s. Nowe wold he haue of eche sort some, to the value of 300 li in the whole, and ech pound one with an other to beare in price 5 s. how muche shalle he haue of eche sorte?

Scholar. That will I trie thus.

[Scholar lays out the problem in the approved manner.]

Mai. I had minded to haue combined them in more varietie, but I am contente to see your owne worke fyrst, and then more varieties in combination may folow anone.

[Scholar works through the problem and arrives at a solution.]

Mayster. Nowe wyll I make the alligation to proue your cunnyng some what better ...

[Master works through the problem another way, arriving at an alternative solution.]

Scho. But if it maye please you to let me see all the variations of this question, before you go from it, for me thinketh I coulde varye it ij.or.iij waies more yet.

Maister. I am content to se you make two or thre variations, but I woulde bee lothe to traye to see all the variations, for it maye be varyed aboue 300 waies, althoughe many of theim wold not well [serve] to this purpose.

Sco. I thought it impossible to make so many variations.

Maist. Meruaile [marvel] not therat, for some questions of this rule may be varied aboue 1000 waies: but I would haue you forget suche fantasies, tyll a tyme of more leysure.

Gentle progress towards the idea of infinity!

That was where Recorde's work ended, his book by now extending to nearly 400 pages. In 1582 a Third Part of some 200 pages was added by John Mellis, easily distinguishable from Recorde's work as it was in Roman type and it abandoned the dialogue form as such, although still setting the work out in a question-and-answer format. Much of it is devoted to the mercantile arithmetic of money, including simple interest, loss and gain, bartering and exchange rates. At the end Mellis adds a short chapter, comprising just six pages, entitled *Sports and Pastimes*. These are all essentially 'think-of-a-number' problems, some of which are rather involved.

Further additions and alterations were made in various subsequent editions, including a section on square and cube roots. The editor of the final edition in 1699, Edward Hatton, condensed everything that had gone before into just 203 pages (admittedly quarto) and added a ninety-three-page supplement on *Decimals Made Easie*. This was a bold attempt to bring the work up to date but it was too late. There was too much competition around by this time, comprising smaller, cheaper works with more exercises. Hatton actually published his own work, *An Intire System of Arithmetic*, in 1721, a quarto volume of 476 pages, pushing the bounds of the subject much wider to include some work on logarithms, algebra, mensuration and simple geometry.

Recorde's Arithmetic was a groundbreaking book. Written in the vernacular, it set out the core syllabus of arithmetic texts for almost the next 300 years. Its dialogue style made the book accessible to the autodidact. That it was appreciated and widely read is evident from the fact that it went through so many editions over such a long period of time. The pedagogical principles it exemplified were sound and were considerably better than those used in a number of its successors. Spiced with wit, it reads agreeably, and is a fine example of what a textbook should be.

Notes

1 Cuthbert Tunstall, *De arte supputandi libri quattuor* (London, 1522).

2 References are to the 1543 edition of *The Ground of Artes* unless specified otherwise.

3 See, for example, Joy B. Easton, 'The early editions of Robert Recorde's *Ground of Artes*', *Isis*, 58/4 (1967), 515–32.

4 The word 'borrow' has ancient origins. Robert Steele (ed.), *The Earliest Arithmetics in English* (London, 1922), quotes from 'The Crafte of Nombrynge', a fifteenth-century manuscript, itself a translation of a work of 1220, which contains sentences such as: 'And then leues 6. cast to 6 the figure of that 2 that stode vnder the hedde of 1. that was borwed & rekened for ten, and that wylle be 8.'

5   See David Eugene Smith, *History of Mathematics* vol. II (1925; Dover edition: New York, 1958), pp. 119–20.

6   Luca Pacioli, *Summa de arithmetica, geometria, proportioni, et proportionalità* (Venice, 1494; 2nd edn Venice, 1523).

7   S. A. Jayawardene, 'Luca Pacioli', in *Dictionary of Scientific Biography* (New York, 1971), pp. 274–7. Moreover, copies are to be found in the major European libraries, and many sixteenth-century works on arithmetic and algebra either mention him or follow his paradigm.

8   For a fuller discussion on Fellowship (also known as Partnership) see Smith, *History of Mathematics*, pp. 554–7; also F. K. C. Rankin, 'The arithmetic and algebra of Luca Pacioli (*c*.1445–1517)' (unpublished PhD thesis, Warburg Institute, University of London, 1992), pp. 239–42.

9   J. M. Pullan, *The History of the Abacus* (London, 1968); F. P. Barnard, *The Casting-Counter and the Counting-Board: A Chapter in the History of Mathematics and Early Arithmetic* (Oxford, 1916).

10   References are to the 1552 edition of *The Ground of Artes*, British Library, STC number 20799.3. A further edition of 1558, Columbia University Library, New York, STC number 20799.5, is based on the 1552 edition.

11   Smith, *History of Mathematics*, pp. 437–41; Rankin, 'Arithmetic and algebra', pp. 229–33.

THREE

Recorde and *The Vrinal of Physick*: context, uroscopy and the practice of medicine

MARGARET PELLING

Iｎ 1935 Sᴀɴꜰᴏʀᴅ Vɪɴᴄᴇɴᴛ Lᴀʀᴋᴇʏ, an American scholar and librarian, jointly authored (with F. R. Johnson) an early work on Recorde.[1] Larkey was unusual for his time in taking a serious interest in lesser-known figures in the history of science and medicine, and in using bibliographical skills – too often taken for granted by other scholars – to reconstruct the routes by which major figures, such as Copernicus and Vesalius, reached a wider audience.[2] Albeit on a lesser scale, Larkey's approach was similar to that of one of the twentieth century's pioneers of the social history of medicine, Henry Sigerist, who was also based at the Medical School and History of Medicine Institute of Johns Hopkins University in Baltimore. Sigerist articulated a number of important principles, including the need to recognize that, in terms of historical events, health precedes disease, and a person might be ill without being treated by medical practitioners. That is, the historian should attend to health-seeking and to the patient, rather than focusing almost exclusively on the medical role. Sigerist also advocated looking at historical individuals in the round and in context, which means, in part, going beyond the very few totemic figures to embrace the intellectual and social environment as a whole.[3]

Recorde is an excellent example of the merits of this approach to the history of medicine. We will not gain a historically accurate idea of the sixteenth century without recognizing that physicians could also be mathematicians, actively involved in mining projects, as well as servants of the Crown in a number of sometimes surprising capacities, including spying and diplomacy. Traditional medical history has not always ignored this kind of diversity, but it has tended to see the respectable aspects of it as polymathic, an indication of the size, scope and refined sensibilities of the medical intellect. While there may be room for admiration along these lines, we need also to recognize that medical practitioners of this period lived a different kind of life from the full-time, vocational, and uniformly educated professionals of the modern day. For those without an assured income from land or the Church, putting together an income meant a constant hunt for opportunities and a similarly constant attention to members of the elite likely to need the skills of those whom we would later call professionals. On a more intellectual level, the sixteenth century witnessed close relationships between fields such as medicine and astronomy, which we would now regard as separate.

All that notwithstanding, it is probably fair to say that Recorde's medical activities, in so far as we can reconstruct them, were the *least* remarkable of his different achievements. However, this does not mean that they are not worth looking at, particularly in relation to other practitioners. Few historians have dealt with Recorde as an individual practitioner, the main exception being Edward Kaplan, who wrote a PhD thesis in 1960 on Recorde as a Tudor scientist, and three years later published an article on Recorde and uroscopy in the *Bulletin of the History of Medicine* (the journal of the Johns Hopkins Institute). Kaplan nevertheless concludes that *The Vrinal of Physick* was exceptional among Recorde's works for an uncritical approach to authority, which could be attributed to the 'primitive condition' of its subject matter. He dismisses medicine in this period as 'one of the most backward areas of science'.[4] The present essay will attempt to consider Recorde's medical identity from a rather different set of perspectives.

It is instructive to compare Recorde with Thomas Linacre, another medical figure of the first half of the sixteenth century. Linacre is better known in history-of-medicine circles than Recorde, primarily as the founder of the London College of Physicians in 1518, but his is still hardly a household name.[5] It would of course be possible to link Recorde profitably with other practitioners who were also mathematicians, notably his contemporary, the physician and cosmographer William Cuningham.[6] However, the comparison with Linacre may bring out more about Recorde's medical characteristics and about medical practitioners of this type in general. Linacre was born about fifty years before Recorde, around 1460, and their similarities and differences reveal much about the changing historical context.

In the first place, both were humanists, trained at Oxford and Fellows of All Souls, and proficient in the classical languages, including the then rare accomplishment of knowing Greek. Subsequently we find a contrast. Linacre travelled to Italy and became one of an international network of humanists including Erasmus and Thomas More. His life's work was the translation into Latin of rediscovered texts of the works of Galen, the second-century Graeco-Roman physician whose synthesis dominated both the theory and practice of medicine for almost 1,500 years. Linacre's translations included Galen's works on the pulse and the temperaments, together with other aspects of treatment and diagnosis. However, Linacre also catered to the contemporary interest in astronomy by publishing a translation of Pseudo-Proclus on the sphere, and he produced elementary textbooks on Latin grammar for the benefit of the newly founded grammar schools as well as in his role as tutor to members of the royal family. The linguistic skills of university-trained physicians, and their humanist convictions about the importance of this form of knowledge, meant that they were likely to involve themselves in educational reform. Linacre published nothing in the vernacular, not even letters.

With respect to another kind of reform typical of the period, that of religion, Linacre was a transitional figure. He was well acquainted with Erasmus, but died in 1524, just before the critical year of Henry VIII's divorce in 1529. Like a number of other physicians, Linacre was an ordained priest and accepted benefices. Because of his pre-Reformation interest in both medicine and religion, he has been adopted as a kind of modern-day icon by medical journals and various other institutions having links with Roman Catholicism.[7] In a rather different spirit, his seamless combination of science, medicine and the humanities made him appear a suitable figurehead for Linacre College, one of Oxford's first graduate colleges, founded in 1962 in the aftermath of the 'two cultures' debate associated with C. P. Snow.

Half a century later than Linacre, Recorde studied very similar texts and acquired similar skills, but applied them on the basis of a very different set of convictions. Whereas Linacre devoted himself to the purification of knowledge, Recorde was concerned with its propagation. Recorde's only published medical work, *The Vrinal of Physick*, his 1547 book on urines, made Galenic medicine available in the vernacular and was his second venture into print. He probably had similar intentions with respect to his other, uncompleted medical project, a work on anatomy. The medical corporation to which he addressed himself in his dedicatory epistle of November 1547 was not Linacre's College of Physicians, but the London Company of Barber-Surgeons, which had been reconstituted as a union of surgeons and barbers under the aegis of Henry VIII in 1540.[8] This did not necessarily mean that Recorde was identifying himself with the barber-surgeons, since physicians claimed the right to supervise the lower ranks of the medical art and to oversee their instruction in anatomy. This is perhaps underlined by the fact that Recorde glosses over the barbers in addressing his epistle 'To the Wardens & company of the Surgians in London' (sig. A.ii). Although some surgeons of this period were highly literate, even in Latin, it would be assumed by those claiming to be their superiors that the barber-surgeons could be instructed only in English.

Recorde's intentions might therefore be described as paternalistic, and as not in any way involving a derogation of his status as a physician.[9] Nonetheless, Recorde did choose to write in the vernacular about the examination of urine, a technique which, if it was respectable at all, belonged to internal medicine and therefore, technically, to physic rather than to surgery. Recorde therefore showed a commitment, not evident in Linacre's work, to the transmission of knowledge on a broad basis, addressing himself, as he put it, 'to all men in commen', 'for the profyte of the hole commens indifferently' (sigs B.iiii–B.iiii[v]). Other Galenic physicians made similar moves at about the same time, notably Linacre's intellectual heir John Caius, credited with re-establishing both the London College of Physicians and Gonville and Caius College in Cambridge. Caius not only published a vernacular tract on the English sweating sickness (1552); he also produced a volume, in Latin, on English dogs, according to the programmes laid down by the humanist encyclopaedists who sought to make new surveys of all aspects of the natural world. The prospective work announced by Recorde in 1551 as 'Of the wonderful works and effects in beasts, plants and minerals' may be seen in this context.[10] It can be presumed that Caius was aiming for a readership, or at least an audience, entirely among the elite. What probably made the difference for Recorde was his firm Protestant convictions and, possibly, his stronger sense of national identity. Caius' reforming tendencies were cautious and relatively short-lived, and he was later accused of crypto-Catholicism.[11]

There is however another kind of resemblance between Linacre, Caius and Recorde in terms of their social identities and in spite of their different religious affiliations. None of them married or had a family. This is not insignificant in the context of the Reformation period. Protestantism, having demoted marriage as a sacrament, tended to elevate marriage and parenthood as socially and morally desirable. Major religious reformers like Zwingli and Luther made a point of marrying and having children, Luther being an ex-monk and his wife an ex-nun. Similarly, but for different reasons, marital status is not incidental to the character of university-trained physicians. In the early modern period, they show a tendency either to marry late or not to marry at all. In Linacre's case of course he could not have married either as a Fellow of a college, or once he had been ordained. Caius and Recorde, on the other hand, are typical of a familial pattern among physicians persisting well into the seventeenth century. This is usually put down to the long years of compulsorily celibate university education undertaken by physicians, and the time it took for them to establish themselves thereafter. While there is some obvious truth in this, as an explanation it is rather anachronistic, and fails to embrace wider issues of gender, social structure and the nature of the work physicians necessarily had to do.[12] Be that as it may, the assumption by later generations that Recorde had an ample number of descendants is as

characteristic of its period as was the fact that he, as an early modern physician, actually had none at all, while making one of his nephews his heir. Family
structure is arguably part of any social group's view of itself, just as suppositions
about fertility are often found in one social group's view of another. Later writers, less likely to explore the family structure of physicians in particular, may have
presumed that Recorde was the head of a family because of his strong Protestant
convictions – or they may simply have been reflecting the assumptions of their
own society.[13]

A similar point can be made in relation to another aspect of social and professional image-creation, and here we find a further resemblance between Linacre
and Recorde. No authentic portrait or sculpted likeness is known to exist of
either Linacre or Recorde, although images of both men were created by later
generations anxious, consciously or otherwise, to fill the gap.[14] For both Linacre
and Recorde, the favoured candidate for a contemporary portrait was one produced by a Flemish artist. This is reasonable in the sense that it was the artists
of northern Europe who were most likely, as Holbein did, to produce portraits
of the mercantile or professional classes. However, notwithstanding the portraits
by Holbein of Henry VIII's physicians John Chambre and Sir William Butts,
it was distinctly *unlikely* that an individual physician would have his portrait
painted at the English courts of the time. This contrasts interestingly with the
greater likelihood that a physician would be recognized in print. As the social
status of physicians increased in later periods, and portrait painting progressed
much further down the social scale, older portraits tended to attract speculative
attributions. For example, any man shown with a skull tended to be identified as
a physician.[15] Ironically, we do have group portraits of the barbers and surgeons,
including one by Holbein produced to mark the occasion of their union by Act
of Parliament under Henry VIII. Physicians lacked, indeed avoided, this kind of
collective identity.[16]

Finally, it is not clear to what extent either Linacre or Recorde actually practised medicine. In each case one needs to take account of the wide range of
their other activities and sources of income, from which we might conclude that
they did not need to practise extensively. However, what we would now regard
as part-time practice was, in fact, characteristic of early modern medicine at all
levels; moreover, given the state of relevant documentation for the period, lack of
evidence proves very little. The sixteenth-century records of the London College
of Physicians, for example, had to be reconstructed retrospectively by John Caius
and are still very meagre for the decades before 1580. (By contrast, the records
of the London Barber-Surgeons' Company, one of the many companies integral
to civic administration in London – as the College was not – show greater con-
tinuity.) The best-known evidence for Recorde's actual practice is the story in

Strype's *Ecclesiastical Memorials* of an incident during the reign of Mary, relating to the Protestant Edward Underhill, known as the 'Hot Gospeller', who was sent to Newgate for writing a ballad containing some flings against Papists. Underhill fell ill in Newgate, as so many did, and Recorde was said to have visited the prison several times to treat him, out of sympathy with Underhill's Protestant convictions and at great risk to himself. Nor did Recorde charge Underhill for his services. This story resembles many such incidents in the martyrologies produced on each side of the religious divide. Physicians were perhaps more likely, instead, to visit prisoners at the behest of the state, as Recorde was said to have done on another occasion recorded by Strype in which he examined, and condemned, a prophesier who had dared to speculate on the life chances of Edward VI. Recorde was chosen for his learning in astronomy and divinity.[17] But this too is a calculated anecdote belonging to what one might call the grand narrative of national affairs. Less noticed, but equally valuable as possible evidence of practice, is Recorde's own remark, made in passing (sigs J.v^v–[J.vi]) in his chapter on 'the commodyties & medicyns of vryne':

> I haue healed with it [men's urine] menny tymes sores on the toes, namely whiche came of bruses, & were without inflammation, & that in seruauntes & husbandmen, which had a ieurney [journey] to go, & no Physition with them / byddyng them to wette a small cloute with it, and to put it into the sores, & then to bynde a cloth about it / & as often as they lysted to make water, to let hyt fall on theire sore toes / & not to take the cloth away tyll hyt were quyte hole.

This brings us back to uroscopy, or the inspection of urines. It was suggested above that this was a technique with highly ambivalent connotations for practitioners, as well as being an essential part of their repertoire. One indication of this ambivalence is the coat of arms chosen for the London College of Physicians.[18] The main feature of the lower half of the arms is a pomegranate, an emblem of fertility, regeneration and resurrection which may seem a surprising symbol for such a homosocial institution, especially because it could also signify the indissolubility of marriage. However, physicians must have hoped to 'resurrect' their patients, and to promote, albeit by proxy, the vital contemporary concerns of human fertility and generation. The pomegranate also bore a particular association with Katherine of Aragon, but the grant of arms was dated 1546, by which time Katherine had been dead for ten years.[19] However it is the diagnostic (or prognostic) signifier shown in the upper half of the arms that we are mainly concerned with here. The urine flask is ubiquitous in western European art and printing as the occupational signifier of the physician.[20] It is depicted, showing calibrations, in Recorde's treatise (see Figure 2). Nonetheless, the College avoided the obvious choice of the urine flask in favour of the pulse. Its coat of arms shows pulse-taking, a hand from above – 'oute of a cloude argent with the Rase [rays] of the Son golde' – lightly clasping the wrist of another hand.[21] As we have seen, Linacre translated Galen on the pulse; he did *not* translate Galen on the urines. This may or may not have been deliberate, but the College of Physicians' choice for its coat of arms appears to be significant in terms of the image it wished to project about itself and its members, and also the image it was seeking to avoid.

The status of physicians in this period was no higher than lower gentry at best.[22] They tended to be the sons of the clergy, parish gentry, or members of the merchant classes. In common with other humanists, however, physicians aspired to join the ranks of those whose incomes derived from who they were, rather than what they did or what they made. These claims to gentle status were not recognized by early modern society, and one of the factors that held physicians back was the nature of their work. Like many of the essential services carried out by women, the practice of physic involved attention to the more repellent aspects of the sick body, including its excretions. Physicians sought to distance them-selves, literally as well as figuratively, from these degrading associations, which is one reason, in my view, why they gave advice by letter or later conducted their practices in coffee houses. The famous William Harvey, when he carried out his duties at St Bartholomew's Hospital in London, attended only once a week. Far from touring the bedsides of his patients, he sat in the main hall while the patients were brought to him.[23] The urine flask admittedly also allowed the phys-ician to maintain this desirable distance from the sick body, as the urine sample was often brought to him by a servant or friend of the patient. Nonetheless, it signified contact with waste matter, and the attempt to avoid such contact, just as much as the contact itself, could turn the physician into an object of ridicule, as we shall see.

Pulse-taking, on the other hand, did not compromise the physician. It involved honourable rather than dishonourable parts of the body, and resembled other forms of polite bodily contact between the higher orders of society. It even carried echoes of the laying-on of hands, or the kind of creative and life-giving benediction famously depicted around 1511 by Michelangelo on the ceiling of the Sistine Chapel, where God the father extends his hand to touch the raised hand of the reclining figure of Adam. As in the Sistine Chapel, the pulse-taking hand in the College's coat of arms comes down from above, from the heavens. Intellectually, the rhythms of the pulse could be linked to the notion of bod-ily harmony as an echo of the harmony of the spheres, as well as to musical theory, one of the liberal arts.[24] The examination of urine could not achieve this degree of elevation.[25] Finally, pulse-taking, unlike uroscopy, represented a form of knowledge that was not readily accessible to outsiders. Urine and its different colours and conditions were visible, the pulse was not. Urine was material, the pulse was not. Detecting different pulses and deciphering their meanings was, and remained, a difficult and obscure art, which physicians could more readily monopolize. The downside was that it was open to sceptics to assert that physicians were inventing phenomena and claiming significances that did not exist. On balance, however, it is easy to see why the College chose pulse-taking rather than the urine flask for its coat of arms. By contrast, the title page of Recorde's *The Vrinal of Physick* (see Figure 1) shows a physician looking at a urine flask raised in one hand, and holding an astrolabe in the other.[26]

Figure 1. Title page of *The Vrinal of Physick* (1547)

At this point, in the cause of balance and historical accuracy, it should be emphasized that the sixteenth century's relationship to urine was different from our own. In developed western societies, urine is merely a waste product which must be made to disappear as soon as hygienically possible. This is true for the urine of domesticated animals as well as of humans. With us, drinking urine is for castaways and celebrity survival shows. We still give samples of urine to doctors for testing, but its nitrogenous content and its shifting character from acid to alkali have little significance for us. For the sixteenth century, on the other hand, urine was both a product and a resource. It was collected and used in a number of manufacturing processes, usually in dyeing or as a bleach. The industrial uses of urine continued into the nineteenth century.[27] Parallels were drawn between urine and wine in terms of their transformations, including fermentation. Like other animal products, including excretions, urine was used as an ingredient in medicines. As such, it was a resource available to the poor.[28] Aristotle, and after him Pliny, thought that one animal, the lynx, hid its urine in the earth because it was so useful; after a certain time, the urine solidified to become amber, a semi-precious substance.[29] Early modern ideas about transmutation meant that any substance had the potential to become other than what it was. Overall, however, although the utility of urine was recognized, those who handled it were traditionally thought of as belonging to dishonourable trades, and physicians shared in this taint.

Uroscopy, or urinoscopy, the examination of urine for assessing state of health or disease, is a very old art, being a feature of both western and eastern ancient civilizations.[30] It did not have a major role in Hippocratic medicine, but was elaborated by Galen. Urine was seen as the waste matter of the food that had been eaten, but also as stemming from all organs, not just the kidneys, so that its condition could provide information about the whole body. By the Middle Ages, urine was being analysed according to a complex system of colours, layers or regions, sediments and suspensions. In order to see these without distortion, the urine flask had to be globular, with no flat base – hence the need for the characteristic basket which also appears in paintings and prints. Vessels of this kind may have been made by Venetian glass blowers from at least the fourteenth century.[31] As many as twenty different colours could be detected in urine. Despite this complexity, writers of the medieval period were already condemning any form of uroscopy practised in isolation from other diagnostic procedures. It is the elevation of uroscopy into a self-sufficient dogmatic system that is condemned as quackery from the Middle Ages into the early modern period, together with the tendency of some uroscopists to distort prognosis into fortune-telling or predicting the future. It did not help that urine inspection could also be practised by women.[32] By the fourteenth century, those who condemned astrology and alchemy might also condemn uroscopy and, by the sixteenth century, we find a flourishing vein of polemic directed against 'piss-prophets'. Thus, uroscopy was respectable if it was hedged about with reservations and used as only one method of diagnosis or prognosis among many. If anyone, man or woman, practised *only* uroscopy, or cried it up as infallible, then he or she was liable to be denounced in medical polemics as a quack.[33]

Quackery was, however, in the eye of the beholder. Simon Forman (1552–1611), the notorious astrological physician of the Elizabethan period also known for his descriptions of Shakespeare's plays and his dreams about Elizabeth I, was regarded by the College of Physicians as a dangerous empiric whose success in London was a threat to its authority. Forman, on the other hand, condemned the standard diagnostic techniques of physicians, including uroscopy, as lazy and misleading:

For they wold mak the pisse and excrement of the bodi to be greter then the bodie yt cam from. Contrary to arte and the true principles of phisicke and philosophie . . . And when they were not able to attain to the trewe science of phisick, which is Astrologie, yt is so depe and tediouse, for that they were lasy and ydelle and not borne to be phisitions but Intruders. Then they began to write bocks of pisspotes and pulses, seges [stools] and other excrements, and so corrupted by tracte of tyme the true science of phisicke, and have filled the wordle with their vain bables and pispot doctrine. By which means they bringe the science of phisicke in question whether yt be a true science or noe. For yf physicke be a science it moste have true and infallible groundes.[34]

In spite of aspiring to be a collegiate physician himself, Forman was reflecting some of the reasons for the relatively low esteem in which orthodox physicians were held because of the nature of their work. In his ambitions for medicine as a science, he resembles an exponent of the seventeenth-century scientific revolution as we now define it, but the means by which he saw medicine achieving this status, through the deep study and application of astrology, were very different. Among the merits of astrology for him was that it could expose deceitful practices in patients, such as lying about their sex lives, or testing the physician by bringing to him the urine of a horse instead of that of the patient.

If we want to see how urine appears in the context of orthodox medical practice of this period, we can look at the *Select Observations on English Bodies*, a casebook put together out of his own practice by Shakespeare's son-in-law John Hall (1574/5?–1635), and published after his death.[35] These cases are not necessarily typical, as they were selected, as stated in the casebook's subtitle, as cases affecting 'eminent persons in desperate diseases', but they are probably the closest we can get to the practice of a physician like Recorde.[36] Evidence of actual practice, as opposed to prescription or polemic where the writer dictates what or what not to do, is often hard to find. Isolated accounts of individual cases can sometimes survive from the sixteenth century, especially in the family papers of the elite, but the only known casebooks, many of them astrological, date from about 1580.[37] Among his cases, Hall includes two that derive from the period 1611–35, where he prescribed the urine of a child as one ingredient in an enema.[38] Sixteenth-century physicians were similarly well aware that urine varied according to age, sex, time of day, amount of sleep and exercise, food eaten and other factors, and this in itself was one reason why Forman thought urine was too deceptive to be an effective diagnostic tool. What is noticeable in Hall's cases is that urine is most often mentioned as a symptom likely to be reported by the patient – for example, blood in the urine after exercise, hot and tormenting urine, and the pain of urine retention in both men and women.[39] This is a reminder that urine and uroscopy belong to the demand as well as the supply side of the medical equation. After many centuries of carrying their urine to a practitioner, patients and their families and friends not surprisingly thought their urine was important and expected the practitioner to be able to decipher it. Both Samuel Pepys and the cleric Ralph Josselin recorded in their diaries that they examined their own urine when ill.[40] Recorde himself urged his readers (sig. B.iiii^v) to be 'diligent to marke their water in tyme of helth' so that they could not only 'enstruct the physicion' but also 'perceyue the cause of the disease sumtymes before the grefe cum'.

Extended information about urine analysis first appeared in print in the English vernacular as an additional section of herbals.[41] With respect to printed vernacular monographs on urines, the *Vrinal* was not the earliest, as it was preceded by at least two anonymous works.[42] Recorde's Preface contains some contemporary references, most notably one to the best-selling health manual by Sir Thomas Elyot, the first extant edition of which is dated 1539. Recorde refers to Sir Thomas as 'that worthye knyght, & lerned clerke' and praises his *Castle of Health* as an English example to match vernacular works on physic being produced in Germany, France and Spain (sig. [B.vii]). In singling out Elyot, Recorde implies that the vernacular was safest in the hands of the learned. On his own account, Elyot showed such promise in learning that before the age of twenty 'a worshypfull phisition, and one of the moste renoumed at that tyme in England' was prompted to read to him works of Galen, together with some aphorisms of Hippocrates. Lehmberg has suggested that the 'worshipful physician' was Thomas Linacre. Like Recorde, but unlike Linacre, Elyot did not feel it necessary to travel to centres of learning abroad. In seeking to disarm the anger of physicians, Elyot asserted that he wrote for their benefit, and in particular 'that the uncertayne tokens of urynes and other excrementes should not deceyve them'.[43]

With respect specifically to the uses of urine, Recorde cites as his authorities Pliny, Dioscorides, Sextus Platonicus (Sextus Papyriensis Placitus), Paul of Aegina, Columella, Constantinus Africanus, Vitalis de Furno, Marcello Virgilio (1464–1521), Ulrich von Hutten (1488–1523), Marsilio Ficino (1433–99) and Galen (sigs J.ii ff.): a mixture of ancient, medieval and modern authors available in printed editions.[44] The first edition of his *Vrinal* of 1547 is black-letter, with a frontispiece and attractive decorated capitals. The Royal College of Physicians' copy of this edition, referred to in this essay, has quite extensive manuscript marginalia, indicating a degree of use or interest on the part of one literate reader at least.[45] But in general, Recorde's book is a rather dense exposition along traditional lines. Although printed marginalia were added to help the reader navigate within the book, it is not well adapted to quick and easy consultation.

Recorde divides his work into six chapters and sets out the content of each chapter at the beginning (sigs C.i–C.i'): firstly, on the nature of urine, 'how it is engendred within man, and how it passes furth from man'; secondly, how to collect it, and 'of the tyme and place meete to consydre it'; thirdly, 'how many thynges ar to be considred in vrin: & how many ways they may be altered in a helthful man'; fourthly, 'what significations & tokens may be gethered of vrine, concernyng any alteration in man, [either] paste, or present, or to cum'; fifthly, 'to what vse in medicine vrine may serue, and of other good vses of it to mannes commoditie'; and lastly, certain diseases affecting the urine, in which urine is either retained or emitted involuntarily, 'with the medicines and remedies meete for the same'. These chapters contrast quite markedly with the kind of ready-reckoner layouts, in which information on colour or regions or prognostications could be quickly located, characteristic of later more popular guides to uroscopy, such as the anonymous *Key to Unknowne Knowledge* of 1599.[46] Indeed, Recorde tried to discourage this kind of readership: the edition of 1547 includes in its preliminaries a doggerel admonishing the reader, printed above a urine flask (complete with regions, see Figure 2) balanced on the ground (sig. A.i'):

> Reade all, or leaue all,
> So am I perfecte and steddye.
> To reade parte and leaue parte,
> Ys to plucke the lymmes from the bodie.

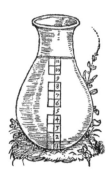

Figure 2. Graduated urine flask in *The Vrinal of Physick*

This applies to the flask of urine, but could equally be meant to apply to how Recorde wanted the reader to use the book itself.

An invaluable study by Paul Slack of sixteenth-century medical literature in the vernacular provides the essential context for Recorde's publication. Of works that survive, about 153 titles, which can be classified as medical, were published in English before 1605. If all known editions are included, this increases to 392. Slack regards this output as modest, concluding that we 'can scarcely attribute any major social or medical impact to a volume of literature of this size'.[47] New works did not exceed reissues until after 1575, and at least one-third of the total were explicitly translations. Of the best-sellers, Slack constructs a list of thirteen that ran to more than six editions before 1605. The book highlighted by Recorde, Elyot's *The Castle of Health*, comes second on this list in terms of editions (sixteen); the anonymous work on urines, *Here beginneth the Seeing of Urines*, which ran to at least thirteen editions between 1525 and 1575, comes fourth, after another early anonymous work, *Here beginneth a New Book of Medicines entitled or called the Treasure of Poor Men* (?1526+) (fifteen editions). At the top of the list is Thomas Moulton's *This is the Mirror or Glass of Health*, first printed before 1531, which appeared in at least seventeen editions between 1530 and 1580.[48] Thus a work on uroscopy does make the top four of the best-seller list for the sixteenth century, but that work is not Recorde's. Indeed the *Vrinal* does not feature at all in Slack's list of the thirteen bestsellers, perhaps underlining the book's lack of 'ready-reckoner' appeal. Slack does conclude that 'the largest sectors of medical publishing in the sixteenth century were . . . the most conservative in quality', suggesting that it was not the *Vrinal*'s lack of originality that restricted its sales.[49]

Overall, Slack finds that orthodox physicians played a relatively small role in producing this literature. Instead, the authors tended to be lawyers, diplomats, civil servants, clergymen or mere scribblers. He sees the likely readership as coming from the same middling social groups.[50] This highlights Recorde's singularity as one of the few physicians active in the vernacular although, as we have seen, it is an underestimate to describe him merely as a physician. He could equally well fit into the category of civil servant. What is notable is that Slack identifies an increase in the number of editions published in the 1540s, and what he calls a 'cluster of productive years' including 1546–8, spanning the last years of Henry VIII's reign and the first of his Protestant son Edward VI.[51] Slack does not explain this particular cluster, and the numbers are small, but Recorde's *The Vrinal of Physick* of 1547 may be seen as part of it. The most obvious explanation, although not necessarily the correct one, is that the transitional period between reigns saw many seeking to recommend themselves and their services to the new regime. Recorde was about thirty-five at this time, but he had only recently moved to London from Cambridge, where he obtained his MD (1545). In career terms, especially for physicians, the *Vrinal* may be seen as a recently qualified man's first major move on the career ladder. In Recorde's case, however, it was not his first publication, and he was aiming at a wider audience. He was also taking certain risks, which probably reflect his strong Protestant convictions.[52] As outlined above, he was giving access to Galenic doctrine on an aspect of physic, in the vernacular, to an audience including not only surgeons, but also the laity and possibly even piss-prophets. He was also committing himself by name to a treatise exclusively on urines at a time when urine-casting, taken singly, had long been suspect, and when fortune-telling and predicting the future were particularly dangerous. In his Preface, he draws an explicit parallel between his own intentions and the teaching of God's word 'vulgarely' to all men, asserting that 'though Cardinalles & monkes haue practised to poyson men euen with the very sacrament of the aulter: yet no man wyll be so mad therfore to eschew the vse of that blessed sacrament' (sigs [B.v]ᵛ–[B.vi]).

Overall, we can conclude that Recorde's multifaceted career, while more remarkable than most, was not in itself unusual among physicians of his type. As the seventeenth century was to show more clearly, physicians could also be natural philosophers, active in the service of the state, or involved in economically significant applications of natural knowledge. Religious and political convictions could be as decisive in the career of an early modern physician as any other factor. Nor can Recorde's medical writings be described as pioneering in their content. Nonetheless, his *The Vrinal of Physick* was notable for its choice of topic, especially as combined with Recorde's use of the vernacular. Recorde may have been seeking to establish himself, but he chose to do so, not by addressing a treatise in Latin to the select few making up the international network of humanist scholars, but by reaching a wider audience that was beginning to define itself, in print, as English.

Notes

1   Francis R. Johnson. and Sanford V. Larkey, 'Robert Recorde's mathematical teaching and the anti-Aristotelian movement', *Huntington Library Bulletin*, 7 (April 1935), 59–87.

2   The author is custodian of a large biographical index of medical practitioners located in London and East Anglia between 1500 and 1700. The index is card-based, with plans to make it available online. It includes cards on Recorde, compiled by the author, mainly from printed sources. It was originally designed as the evidence base for a project leading to the publication of the collected volume by Charles Webster (ed.), *Health, Medicine and Mortality in the Sixteenth Century* (Cambridge, 1979), dedicated to the memory of Sanford Vincent Larkey. This volume contains a bibliography of Larkey's works.

3   On Henry Sigerist see Genevieve Miller (ed.), *A Bibliography of the Writings of Henry E. Sigerist* (Montreal, 1966); Elizabeth Fee and Theodore M. Brown (eds), *Making Medical History: The Life and Times of Henry E. Sigerist* (Baltimore, 1997).

4   Edward Kaplan, 'Robert Recorde and the authorities of uroscopy', *Bulletin of the History of Medicine*, 37 (1963), 65–71, esp. pp. 69, 71; *idem*, 'Robert Recorde, *c*.1510–1558: studies in the life and work of a Tudor scientist' (unpublished PhD thesis, New York University, 1960). I have not been able to consult Kaplan's thesis.

5   For what follows on Linacre, see Francis Maddison, Margaret Pelling and Charles Webster (eds), *Linacre Studies: Essays on the Life and Work of Thomas Linacre c.1460–1524* (Oxford, 1977).

6   Louise Diehl Patterson, 'Recorde's cosmography, 1556', *Isis*, 42 (1951), 208–18, esp. p. 209; Webster, *Health, Medicine and Mortality*, p. 203.

7   See Margaret Pelling, 'Published references to Thomas Linacre', in Maddison, Pelling and Webster, *Linacre Studies*, pp. 343ff.

8   As a craft guild of barbers, the London Barber-Surgeons' Company dated from at least 1308. The surgeons finally split from the barbers in 1745 and formed a separate company. See Sidney Young, *The Annals of the Barber-Surgeons of London* (London, 1890; repr. New York, 1978).

9   This is the line taken on the *Vrinal* by Samuel Lilly, 'Robert Recorde and the idea of progress: a hypothesis and [a] verification', *Renaissance and Modern Studies*, 2 (1958), 3–37. Lilly posits a complete change in Recorde's approach between 1547 and 1551, but it is also worth noting Lilly's low estimate of both surgeons and guilds.

10   See Stephen Johnston, 'Recorde, Robert (*c*.1512–1558)', *Oxford Dictionary of National Biography* (Oxford, 2004), online edn accessed 21.11.09.

11   See John Caius, *A Boke or Counseill against the Disease called the Sweate* (London, 1552; facsimile edited by Archibald Malloch, New York, 1937); Vivian Nutton, 'John Caius and the Linacre tradition', *Medical History*, 23 (1979), 373–91; Nutton, *John Caius and the Manuscripts of Galen* (Cambridge: Cambridge Philological Society suppl. vol. 13, 1987); and Nutton, 'Caius, John (1510–1573)', *Oxford Dictionary of National Biography* (Oxford, 2004), online edn accessed 23.08.11. For recent discussions of the sweating sickness, see Alan Dyer, 'The English sweating sickness of 1551: an epidemic anatomized', *Medical History*, 41 (1997), 362–84; Danae Tankard, 'Protestantism, the Johnson family and the 1551 sweat in London', *London Journal*, 29 (2004), 1–16.

12   For discussion of these issues see Margaret Pelling, 'The women of the family? Speculations around early modern British physicians', *Social History of Medicine*, 8 (1995), 383–401.

13   Simple error is, of course, also a possibility. See the correction of the *Dictionary of National Biography* entry of 1896 in David Eugene Smith and Frances Marguerite Clarke, 'New light on Robert Recorde', *Isis*, 8/1 (1926), p. 52. Linacre was also wrongly credited with a wife, *c*.1631: Maddison, Pelling and Webster, *Linacre Studies*, p. xxxiv.

14   See Maureen Hill, 'An iconography of Thomas Linacre', in Maddison, Pelling and Webster, *Linacre Studies*, pp. 354–74; J. W. S. Cassels, 'Is this a Recorde?' *Mathematical Gazette*, 60/411 (March 1976), 59–61.

15   As was the case with Linacre: see Hill, 'An iconography of Thomas Linacre', pp. 362, 371–2, and Plate XIV.

16   See Young, *Annals of the Barber-Surgeons*, pp. 80–94. The Holbein painting, which exists in two versions, also conveys the medical hierarchy at court by including two of the royal physicians, John Chambre and William Butts, and the royal apothecary, William Alsop. See also Roy C. Strong, 'Holbein's cartoon for the barber-surgeons' group rediscovered – a preliminary report', *Burlington Magazine*, 105 (1963), 4–14; Bertram Cohen, 'A tale of two paintings', *Annals of the Royal College of Surgeons of England*, 64 (1982), 4–12. A later collective portrait shows the surgeon John Banister lecturing on anatomy at Barber-Surgeons' Hall, London in 1581: see D'Arcy Power, *Notes on Early Portraits of John Banister, of William Harvey, and the Barber-Surgeons' Visceral Lecture in 1581* (London, 1912). For the College and portraits, see Margaret Pelling, *Medical Conflicts in Early Modern London: Patronage, Physicians, and Irregular Practitioners 1550–1640* (Oxford, 2003), p. 58; on collective identity see pp. 20–1 and *passim*.

17   John Strype, *Ecclesiastical Memorials; Relating chiefly to Religion, and the Reformation of it*, 3 vols (London, 1721), vol. II, p. 113; vol. III, p. 64. The account is Underhill's own, as given to John Foxe. For these anecdotes see also Smith and Clarke, 'New light on Robert Recorde'. For Underhill see John N. King, 'Underhill, Edward (b. 1512, d. in or after 1576)', *Oxford Dictionary of National Biography* (Oxford, 2004), online edn accessed 23.08.11.

18   George Clark, *A History of the Royal College of Physicians of London*, 2 vols (Oxford, 1964–6), vol. I, Plate III; A. Stuart Mason, 'The arms of the College', *Journal of the Royal College of Physicians*, 26 (1992), 231–2. Mason notes that the hand, in modern terms, is wrongly positioned for pulse-taking. There seems to be no extant contemporary record of how or why the arms were chosen by the College.

19   Patricia Langley, 'Why a pomegranate?' *British Medical Journal*, 321 (4 November 2000), 1153–4. Langley lists medicinal uses for pomegranates, dating from Dioscorides.

20   See Joseph H. Kiefer, 'Uroscopy: the artist's portrayal of the physician', *Bulletin of the New York Academy of Medicine*, 40 (1964), 759–66.

21   Mason, 'The arms of the College', quoting the heraldic description in the grant of 1546.

22   On what follows, see Pelling, *Medical Conflicts in Early Modern London*; eadem, 'Compromised by gender: the role of the male medical practitioner in early modern England', in Hilary Marland and Margaret Pelling (eds), *The Task of Healing: Medicine, Religion and Gender in England and the Netherlands 1450–1800* (Rotterdam, 1996), pp. 101–33.

23   D'Arcy Power, *William Harvey* (London, 1907), pp. 35–7.

24   Nancy G. Siraisi, 'The music of pulse in the writings of Italian academic physicians (fourteenth and fifteenth centuries)', *Speculum*, 50 (1975), 689–710.

25   Cf. Faith Wallis, 'Signs and senses: diagnosis and prognosis in early medieval pulse and urine texts', *Social History of Medicine*, 13 (2000), 265–78.

26   I am grateful to Stephen Johnston for his advice as to the astrolabe.

27   See John Burnett, 'William Prout and the urinometer: some interpretations', in Robert G. W. Anderson, James A. Bennett and William F. Ryan (eds), *Making Instruments*

Count: Essays on Historical Scientific Instruments (Aldershot, 1993), p. 246, and J. R. Partington, *A History of Chemistry*, 4 vols (London, 1961–70).

28 For example, as a preservative against the plague: Andrew Wear, *Knowledge and Practice in English Medicine, 1550–1680* (Cambridge, 2000), p. 337. This is mentioned by Recorde as a practice known in Syria: *Vrinal*, sig. [J.vi].

29 Partington, *A History of Chemistry*, vol. I, pt 1, p. 112.

30 See V. Aalkjaer, 'Uroscopia: a historical and art historical essay', *Acta Chirurgica Scandinavica*, Suppl., 433 (1973), 3–11; Partington, *A History of Chemistry*; L. Y. Baird, 'The physician's "urinals and jurdones": urine and uroscopy in medieval medicine and literature', *Fifteenth-Century Studies*, 2 (1979), 1–8.

31 Nancy G. Siraisi, *Medieval and Early Renaissance Medicine: An Introduction to Knowledge and Practice* (Chicago, 1990), p. 144.

32 Such practice by women was not necessarily illicit: see ibid., p. 34.

33 See for example Vivian Nutton, 'Idle old trots, coblers and costardmongers: Pieter van Foreest on quackery', in Henriette A. Bosman-Jelgersma (ed.), *Petrus Forestus Medicus* (Amsterdam, 1997), pp. 245–54. James Hart's translated and abridged version of Foreest's polemic was published as *The Arraignement of Urines* in 1623.

34 Simon Forman, quoted in Lauren Kassell, *Medicine and Magic in Elizabethan London. Simon Forman: Astrologer, Alchemist, and Physician* (Oxford, 2005), p. 119.

35 Hall's casebook is reproduced in facsimile and with commentary in Joan Lane (ed.), *John Hall and his Patients: The Medical Practice of Shakespeare's Son-in-Law* (Stratford-upon-Avon, 1996). It was first published in 1657, as a translation from Hall's original Latin.

36 Although it seems accurate to describe Hall as an orthodox physician, it should be noted that nothing is known for certain of his education after his MA from Cambridge in 1597. See Joan Lane, 'Hall, John (1574/5?–1635)', *Oxford Dictionary of National Biography* (Oxford, 2004), online edn accessed 23.08.11.

37 See Appendix I, 'English casebooks, 1500–1700', in Lauren Kassell, 'Simon Forman's philosophy of medicine: medicine, astrology and alchemy in London, c.1580–1611' (unpublished DPhil thesis, Oxford University, 1997), pp. 221–2.

38 Lane, *John Hall and his Patients*, pp. 326–8. For comments by Recorde on children's urine, see *Vrinal*, sigs J.iii, [J.vi].

39 Lane, *John Hall and his Patients*, pp. 4, 7, 96–7, 213–15, 222, 232–3, 242. For an isolated indication that Hall inspected urine, see p. 244: 'the urine was confused, and there was great danger of death.' For a similar concern about confused urine, see Lucinda McCray Beier, *Sufferers and Healers: The Experience of Illness in Seventeenth-Century England* (London, 1987), p. 104.

40 Beier, *Sufferers and Healers*, p. 23.

41 As, for example, in *The Great Herbal Newly Corrected* of 1539. This herbal was first printed in English in 1526. See Agnes Arber, *Herbals, their Origin and Evolution*, 3rd edn

(Cambridge, 1986), pp. 44–50, 274. Recorde cites as a positive precedent the appearance of herbals in the 'vulgare tong': *Vrinal*, sig. [B.vii].

42   Namely, *Here beginneth the Seeing of Urines*, first printed for Richard Banckes in 1525 and in numerous editions thereafter; and *The Judycyall of Uryns* (Southwark, 1527?). Postdating Recorde's treatise were *Hereafter foloweth the Judgement of all Urynes* (London, 1555?), and Humphrey Lloyd's translation of Jean Vassès, *Here beginnith a litel treatise conteyninge the jugement of urynes* (London, 1553). Titles of early works have been modernized unless a specific edition is cited.

43   Sir Thomas Elyot, *The Castel of Helth Corrected* (London, 1541), sig. [A4]. See Stanford Lehmberg, 'Elyot, Sir Thomas (*c*.1490–1546)', *Oxford Dictionary of National Biography* (Oxford, 2004), online edn accessed 23.08.11. Elyot's 'Proheme' to the 1541 edition, which replaces the dedication to Thomas Cromwell of the first edition, opens with praise of Galen.

44   For Recorde's list of authorities for the treatise as a whole, where he foregrounds the ancients by way of justification, see *Vrinal*, sig. B.ii', and Kaplan, 'Robert Recorde and the authorities of uroscopy'. Columella was a first-century writer on agriculture and gardens; it is perhaps more surprising that Linacre owned a work by this author (Maddison, Pelling and Webster, *Linacre Studies*, p. 333).

45   At least one owner can be identified. The seventeenth-century catalogue of the private library of the Le Stranges of Hunstanton, Norfolk lists the *Vrinal* (stated to be an edn of 1597, not otherwise known), as well as Newton's translation of Levinus Lemnius's *Herbal* and the best-sellers, Moulton's *Mirror* and Elyot's *Castle*. I am grateful to Jane Whittle for access to this information in advance of the publication of her book (with Elizabeth Griffiths), *Consumption and Gender in the Early Seventeenth-Century Household: the World of Alice Le Strange* (Oxford, 2012).

46   *The Key to Unknowne Knowledge. Or, a Shop of five Windowes . . . Consisting of five necessarie Treatises: Namely, 1. The Judgement of Urines 2. Judiciall rules of Physike . . .* (London, 1599).

47   Paul Slack, 'Mirrors of health and treasures of poor men: the uses of the vernacular medical literature of Tudor England', in Webster, *Health, Medicine and Mortality*, pp. 237–73: pp. 238–40. The proportion of medical works in English publishing overall is estimated as around 3 per cent (ibid., pp. 239–40).

48   Ibid., pp. 240, 242, 247–8.

49   Ibid., pp. 251, 248. Some allowance might also be made for the fact that Recorde's work appeared not in the first quarter of the century, but near the end of the second.

50   Ibid., pp. 252–4, 256.

51   Ibid., pp. 240–1.

52   For Recorde's expression of 'commonwealth sentiments' in 1552, see Joy B. Easton, 'The early editions of Robert Recorde's *Ground of Artes*', *Isis*, 58/4 (1967), 515–32.

FOUR

The Pathway to Knowledg and the English Euclidean tradition

JACQUELINE STEDALL

R OBERT RECORDE'S *The Pathway to Knowledg*, the second of his mathematical
textbooks, was published in 1551, some eight years after *The Ground of Artes*.
Its metaphorical title reinforces the trope of mathematical discovery as a journey,
an adventure even: the student well grounded in arithmetic was now ready to set
out in a new direction, towards mastery of geometry. The *Pathway* was the first
textbook of geometry for English readers, and for that reason alone is historically
fascinating. It can also be seen from a different perspective, however, as one of
many texts in a long line of translations, editions or other renderings of Euclid's
Elements, first written in Greek around 250 BC. Recorde was neither the first nor
the last to bring the *Elements* to England, though he was the first to do so in his
native language. This chapter examines the contents of the *Pathway* in the context
of its time but also as part of a tradition that extends down to the present, of
continually reinventing Euclid for new generations of learners.

The *Elements* before Recorde

Euclid's *Elements* can be thought of as an encyclopaedia of geometry in thirteen
books, a systematic collection of problems and theorems based on the properties
of straight lines, angles and circles. There are no contemporary references to the
Elements in antiquity and we know nothing at all about Euclid himself, but some-
how his work survived to become the most widely used textbook of all time. By
the seventeenth century Euclid's *Elements* were known from England to China, and

continued to be part of the school curriculum in Britain and elsewhere in Europe until the twentieth century.

For centuries, however, the preservation and survival of the *Elements*, as for any Greek text, were very much a matter of chance. Written on fragile materials like papyrus, the original and all early copies have long since vanished. In the Latin West, a few easy propositions were preserved in the writings of the Roman writer Boethius (*c.*AD 500) and in most of Europe for hundreds of years these were all that were known. Fortunately the situation elsewhere was better. Full copies of the *Elements* were preserved, for instance, in the Greek-speaking parts of the old Roman empire: the oldest extant copy, now in the Bodleian Library in Oxford, was created in Byzantium in AD 888. Meanwhile many translations were made from Greek to Arabic, with accompanying commentaries, between about AD 750 and 1250. The first medieval translations into Latin were thus from Arabic rather than Greek, some of them made by such English scholars as Adelard of Bath (1130) or Robert of Chester (1145). During the twelfth and thirteenth centuries, manuscript copies of the *Elements* in Latin began to find their way into monastic libraries or the fledgeling universities of Paris, Oxford and Cambridge, but would have been seen and read only by the very few monks or scholars who had access to them.

The first printed edition of the *Elements*, based on the thirteenth-century Latin translation by Campanus of Novara, was published in Venice in 1482 by Erhard Ratdolt.[1] A few years later, Bartolomeo Zamberti, scathingly critical of some of the 'barbarous' (Arabic) terms retained by Campanus, published his own Latin translation, made directly from Greek (Venice, 1505). Four years later, Luca Pacioli revised Ratdolt's Latin edition in support of Campanus (Venice, 1509). The translations of both Campanus and Zamberti were later republished in a dual edition by Jacques Lefèvre d'Étaples in Paris in 1516. The first Greek edition was published by Simon Grynaeus in Basel in 1533. Many of these editions were beautifully produced, and therefore expensive. All required a knowledge of classical languages: there was no edition in any vernacular language until an Italian translation by Niccolò Tartaglia was published in Venice in 1543.

In the light of this brief history we can begin to understand the novelty and the challenge of what Recorde aimed to do: to make some of the geometry of the *Elements* available to ordinary Englishmen in their own language. His task was a double one, neither part of it easy. First, he had to translate Greek and Latin terms for 'triangle', 'trapezium', 'acute angle' and many others, into a language that had no natural equivalents (a modern comparison might be writing a computer manual in Gaelic or Finnish). Either he had to carry over Latin words into English, or he had to invent new English words to carry old meanings. Recorde's deep love for and mastery of the English language persuaded him to the latter. His second challenge was more profound: to introduce geometry itself to a population that had hardly heard of such a subject, let alone studied it. Tradesmen and craftsmen of course knew how to apply technical procedures, but these had been passed on by hand and by word of mouth from master to apprentice for generations without any need for book learning. What then was this subject called 'geometry'? What was it about? What was the use of it? It is almost impossible for us now to imagine the newness and strangeness of the material that Recorde was offering, but it is only by bearing it in mind that we can begin to understand his approach and his choice of examples.

As for Recorde's sources, we can gather some information about them not only from his direct acknowledgements but also from some of the unusual terms and diagrams he used in the *Pathway*. Early in the text, in connection with the construction of parallel lines, he mentions Albrecht Dürer (sig. [A.iiij]^v), whose treatment of Euclidean geometry and perspective had been published in *Underweysung der Messung* in 1525. Further on, he cites both Boethius and Georg Joachim Rheticus as writers who had set out many of Euclid's propositions, though without demonstrations (sig. a.ij^v). A treatise on geometry attributed to Boethius was published in Venice in a collection entitled *Hec sunt opera Boetii* in 1491. Rheticus, however, was a much more recent writer: his *De lateribus et angulis triangulorum* was published in 1542.

Jack Williams has traced some of the more unusual words used by Recorde in his initial definitions: 'isopleuron' (equilateral) and 'aequicurio' (isosceles) uniquely to Martianus Capella, a fifth-century Roman writer whose works were first printed in Italy in 1499; 'mensula' (trapezium) to the supposed geometry of Boethius first printed in 1491; and 'helmeariphe' (trapezium) to Campanus.[2] The last suggests that Recorde was familiar with at least one of the early Venetian editions of the *Elements*, by Ratdolt, Zamberti or Pacioli. The fact that some of Recorde's diagrams are very like those in Jacques Lefèvre d'Étaples's Campanus–Zamberti edition of 1516 (reprinted in Basel in 1537) argues that he may well have drawn on that text in particular. There is also a suggestion (detailed below) that Recorde knew the 1533 Grynaeus Greek edition. Recorde could have found these books in the libraries of Oxford or Cambridge colleges (he was a Fellow of All Souls College, Oxford from 1531 but moved to Cambridge in the early 1540s).[3] He might also have owned his own copies, because we know that he acquired recent books from the Continent through his friend Reynold Wolfe, the London printer who published all his earlier mathematical texts.

Thus, a few years after the publication of *The Ground of Artes*, working from some of the most comprehensive texts then available, Recorde began work on the book that was eventually published as *The Pathway to Knowledg*.

Euclid's *Elements*, Books I to IV

In this chapter we shall be concerned only with the first four books of the *Elements*, but for completeness here is a very brief overview of all thirteen. Books I to VI of the *Elements* treat plane geometry: the properties of points, lines, angles, circles and polygons, ratios of magnitudes, and similar figures. Books VII to X change course to discuss properties of numbers: whether they are odd, even, square, triangular, prime, perfect and so on, though all of this is still couched in geometrical language. Books XI to XIII return to geometry proper, this time solid geometry, ending with constructions of the five regular solids: the cube, tetrahedron, octahedron, icosahedron and dodecahedron. Recorde wrote about some of the more advanced ideas from Book V onwards in *The Whetstone of Witte* in 1557, but in the *Pathway* he focused only on Books I to IV.

Book I opens with three special types of statement: (i) *definitions*: what we start from; (ii) *postulates*: what we are allowed to do; (iii) *common notions*: what we can all agree upon. These are then followed by a series of forty-eight carefully constructed propositions, each built only on what has gone before.

Euclid's first definition is of a point: 'A *point* is that which has no part,' followed by 'A *line* is a breadthless length.'[4] One can argue about these, of course: what did Euclid mean by the negative descriptions 'has no part' or 'breadthless' when he had not defined either 'part' or 'breadth'? The problem is that one has to start somewhere, and the more basic the 'element' one is trying to define, the harder it is to know where to begin. Most readers can probably live with Euclid's definitions without too much unhappiness. When he goes on to tell us that 'The extremities of a line are points' we then understand that by a 'line' he means a finite length, or what we might call a line segment, rather than an infinite line extending out into space. Next comes another idea that is easy to understand intuitively but which is very difficult to pin down precisely: 'A *straight* line is a line which lies evenly with the points on itself'. 'Lies evenly'? What does that mean? Again, we simply have to accept from our everyday experience what Euclid is trying to say, and move on.

The next few definitions, if we have accepted the first few, are less difficult: 'A *surface* is that which has length and breadth only'; 'A *plane angle* is the inclination to one another of two lines in a plane which meet one another'; and so on. Subsequent definitions give us 'a right angle', 'a perpendicular', 'acute' and 'obtuse' angles, 'a figure', 'a circle', 'a diameter', various kinds of 'trilateral' and 'quadrilateral' figures, and 'parallel' straight lines.

Next come the postulates, which specify the actions we are allowed to perform. There are essentially just three of them, namely: (i) we may draw a straight line from any point to any point; (ii) we may produce (that is, extend) a finite straight line continuously in a straight line; (iii) we may draw a circle with any centre and radius. These are the rules that restrict Euclidean geometry to constructions with an unmarked ruler and compass only. No absolute measurement enters into any of this: the radius of a circle, for instance, is simply a length that is copied by compass from some other line: lengths are only relative to each other. The fourth postulate appears misplaced since it seems to be not something one performs but something one has to agree with, namely, that all right angles are equal to one another; however, it can be taken to mean that one can compare figures drawn in different positions or orientations. The fifth postulate later became the most troublesome of all. It states that if two straight lines are cut by a third line, then the first two will meet, and will do so on the side where the angles made by the cutting line are less than two right angles. Many later commentators thought that this should be a theorem rather than a postulate, except that proving it turned out to be extraordinarily difficult. This problem was not finally resolved until the nineteenth century with the discovery of non-Euclidean geometries, but that takes us far beyond our present exploration.

Finally, the introductory material to Book I gives the common notions, ideas that one can hardly disagree with, for example: 'Things that are equal to the same thing are also equal to one other', 'If equals be added to equals, the wholes are equal', and 'The whole is greater than the part'.

Only after this careful setting up of his system did Euclid begin his propositions, each of them stated and proved from what had gone before. Some of them are what were later called 'problems', requiring the reader (either in the mind or in practice) to do or construct something. Proposition I.1, for instance, teaches the construction of an equilateral triangle on any given line segment *AB*. Euclid first gives his instructions: draw a circle with centre *A* and radius *AB*; repeat at *B*; take a point *C* where the two circles cross and join it to both *A* and *B* to produce triangle *ABC*. Clearly this construction is possible within Euclid's framework: it requires no more than the definitions of lines and circles together with the first and third postulates. Euclid did more, however, than just offer a construction: he also proved carefully and rigorously that the triangle so created is indeed equilateral.

Other propositions are not 'problems', to be solved by construction, but 'theorems', which require a theoretical deduction. Proposition I.4, for example, claims that if two triangles are equal with respect to two of their sides and the angle between them, they will also be equal in every other respect. No construction is needed here; instead the proof can be given as a sequence of logical steps.

Many familiar procedures and facts from school geometry are to be found in the remainder of Book I: Proposition I.9 shows how to bisect any given angle; Proposition I.32 proves that angles in a triangle always add up to two right angles;

and finally, in Proposition I.47, we have a proof of what is universally known as 'Pythagoras' theorem'. In keeping with the sprit of the *Elements* it appears as a purely geometric theorem, about squares that are geometrically constructed on the three sides of a right-angled triangle.

Book II compares the sizes of various quadrilateral figures. Proposition II.4, for example, states that if a straight line is cut at random, the square on the whole line is equal to the sum of the squares on the two segments together with twice the rectangle formed by the segments. If we imagine representing the two line segments by lengths a and b we can write this proposition easily enough as $(a + b)^2 = a^2 + b^2 + 2ab$ and assent to the truth of it. This has led some commentators to suggest that Euclid was really doing a kind of 'geometric algebra', but that fails to do justice to the spirit of his book as a whole and to this particular proposition, which is written and proved in purely geometric terms.

Book III treats circles in the same way that Books I and II treated angles, triangles and quadrilaterals, beginning with definitions of 'equal circles', 'touching circles', 'segments' and 'sectors'. The remainder of Book III contains the circle theorems that many people will remember from school geometry: that angles in the same segment are equal, for instance, and many others of a similar kind.

Finally, Book IV deals with inscribed and circumscribed figures. Several propositions show how to inscribe various kinds of triangle inside a circle, or a circle inside a given triangle. There are then further constructions for squares, regular pentagons and regular hexagons, and in Proposition IV.16, the last, for a regular fifteen-sided figure inside a circle. Such constructions are required later for Euclid's ultimate goal of constructing the regular solids, but there is no solid geometry in Book IV itself.

With this overview of Books I to IV of the *Elements* in mind, let us now turn to Recorde's *Pathway*.

The Pathway to Knowledg (1551)

The title page of the *Pathway* clearly announces Recorde's intention of providing practical geometrical knowledge for use in everyday life. Like many sixteenth-century book titles, Recorde's includes a description of the subject matter: 'The pathway to knowledg, containing the first principles of Geometrie, as they may moste aptly be applied vnto practise, bothe for vse of instrumentes Geometricall, and astronomicall and also for proiection of plattes in euerye kinde, and therfore much necessary for all sortes of men'. Lower down the page, however, we read another message, a subtext perhaps, of the main story. It is presented in the kind of short verse that Recorde was fond of inserting into all his texts; this one is entitled 'Geometries verdicte' and reads:

> All fresshe fine wittes by me are filed,
> All grosse dulle wittes wishe me exiled:
>   Thoughe no mannes witte reiect will I,
> Yet as they be, I will them trye.

Here then, is geometry not for practical purposes, but in a more traditional guise, a pursuit that will exercise and sharpen the mind. These two aspects of geometry, the practical and the theoretical, are to be held in a healthy tension by Recorde, not just on this first page but throughout the book.

The list of contents on the back of the title page shows that the *Pathway* was originally conceived as four books, but that only the first two were published, the two that display the strongest Euclidean influence. The first contains 'definitions of the termes and names vsed in Geometry' and the second 'the Theoremes'. The third and fourth books appear to have been written but not printed for reasons of 'other hindrances', and it is not altogether clear what they would have contained. The third promised treatment of 'diuers formes, and sondry protractions [projections?] thereto belonging', possibly a treatment of perspective, while the fourth was to teach 'the right order of measuringe'.

There are several preliminary pieces of varying length in Recorde's text before we come to any geometry. The first is a two-page note 'to the gentle reader' (sigs r.ii–r.ii^v). It begins with the following passage:

> Excvse me, gentle reader if oughte be amisse, straung paths ar not troden al truly at the first: the way muste needes be comberous, wher none hathe gone before. where no man hathe geuen light, lighte is it to offend, but when the light is shewed ones, light is it to amende. If my light may so light some other, to espie and marke my faultes, I wish it may so lighten them, that they may voide offence.[5]

This reads at first sight as a lovely and intricate piece of English prose, with multiple play on the word 'light'. It does not take long, however, to realize that it is actually in rhyme. From time to time, Recorde's prose takes flight and transforms itself into poetry:

> Excvse me, gentle reader if oughte be amisse,
> straung paths ar not troden al truly at the first:
> the way muste needes be comberous, wher none hathe gone before.

where no man hathe geuen light, lighte is it to offend,
but when the light is shewed ones, light is it to amende.
If my light may so light some other, to espie and marke my faultes,
I wish it may so lighten them, that they may voide offence.

Again and again Recorde returned to the metaphor of light: 'This candle did I light: this lighte haue I kindeled: that learned men maie se, to practise their pennes . . . that finer wittes maie fashion themselues with such glimsinge dull light, a more complete woorke.' Finally, Recorde made a plea of a kind he often repeated: 'And this gentle reader I hartelie protest where erroure hathe happened I wisshe it redrest.' Once again his prose disguises a rhyming couplet.

After this short note to the reader comes a much longer and more formal letter to 'the most noble and puissaunt prince Edwarde the sixte' (sigs r.iii–s.iii^v). When the *Pathway* was published, Prince Edward was fourteen years old and had already been king of England for four years. Recorde's letter to Edward is quite different in tone from his address to the common reader, now more like a pious sermon. He argues that all men are agreed on two things, the pursuit of happiness and the pursuit of power. He expands at some length on the latter, somewhat disingenuously venturing to suggest that 'bokes dare speake, when men feare to displease.' His theme is that power may be combined with royalty or with wisdom, and that, as the examples of Alexander and Solomon demonstrate, knowledge is as valuable as power. The best combination of attributes, however, was demonstrated by Solomon's father David, who displayed both wisdom and virtue. Further, such wisdom must be in both religious and human affairs; and as far as the latter are concerned the mathematical sciences, especially arithmetic and geometry, must be foremost. Earlier in this long letter, Recorde had mentioned Alexander's good fortune to be born in the time of Aristotle; now he points out that Prince Edward himself lives in an age of 'skilful schoolmaisters & learned techers', amongst whom he clearly counts himself since a few lines earlier he had offered to write textbooks either in Latin or English for the universities of Oxford and Cambridge (an offer that was never taken up).

This is still not the end of the preliminaries. Next comes a Preface, once again written for the common reader (sigs [s.iiii]–t.iii^v). The text moves out of the tightly printed gothic font of the address to Prince Edward and returns to the softer italic of the note to the 'gentle reader'. The heading is: 'The preface declaring briefely the commodities of Geometrye, and the necessitye thereof'. It was conventional, of course, for writers to assert the value and importance of the discipline they were promoting, but Recorde was unusual in the thoroughness of his claims. His Preface describes in considerable detail the many classes of men for whom geometry is useful. Recorde began with 'the vnlearned sorte', who need geometry in 'measuryng of ground, for medow, corne, and wodde: in hedgyng, in dichyng, and in stackes makyng'. Though the husbandman digging a ditch may not acknowledge the rules of geometry, for example, he cannot succeed unless 'he kepe not a proportion of bredth in the mouthe, to the bredthe of the bottome, and iuste slopenesse in the sides agreable to them bothe'.

As for 'Carpenters, Karuers, Joyners, and Masons', and a multitude of other craftsmen, geometry is crucial. Indeed it is so common and all-pervasive in their work that Recorde once again drifts into the easy rhythm of verse:

> The Shippes on the sea with Saile and with Ore,
>   were firste founde, and styll made, by Geometries lore,
> Their Compas, their Carde, their Pulleis, their Ankers,
>   were founde by the skill of witty Geometers.
> To sette forth the Capstocke, and eche other parte,
>   wold make a great showe of Geometries arte.
> Carpenters, Caruers, Joiners and Masons,
>   Painters and Limners with suche occupations,
> Broderers, Goldesmithes, if they be cunning,
>   Must yelde to Geometrye thanks for their learning.

And so on.

The 'learned professions', however, are treated more formally in prose. Recorde appealed to Aristotle to support his claim that geometry is required in logic, rhetoric and philosophy; to Galen and Hippocrates for its usefulness in medicine; to Plato, Aristotle and Lycurgus for its necessity in law-making; and to Archimedes for proof of its help in war. He also claims that geometry is helpful in the study of divinity but has little actually to say about this except that 'I shoulde seeme somewhat to tedious, if I shoulde recken vp, howe the diuines also in all their mysteries of scripture doo vse healpe of geometrie.' Towards the end of his Preface, however, he is at pains to point out how seemingly magical or unnatural occurrences can often be understood with the help of geometry: a machine invented by Archimedes to shoot one hundred darts, for instance, led to tales of a giant with one hundred hands; while Bacon's reflecting glass led men to believe in a mirror in which 'men myght see thynges that were doon in other places, and that was iudged to be done by power of euyll spirites'. Contrary to popular belief, claims Recorde, there is no magic in any of this because 'the reason of it [is] good and naturall, and to be wrought by geometrie'. Here then is a more subtle use of geometry than any of those he has described so far: to banish ignorance and superstition.

But now to the geometry itself (sigs A[.i]–[I.ii]). After a brief statement that geometry is concerned with the drawing, measuring and proportion of figures Recorde begins, like Euclid, with definitions of a *point*, a *line*, and a *straight* line. At first, he seems to be simply repeating Euclid's text: 'A *Poynt or a Prycke*, is named of Geometricians that small and vnsensible shape, whiche hath in it no partes, that is to say: nother length, breadth nor depth.' Almost immediately, however, he digresses to his own more practical definition, much better suited to his likely readership: 'I thynke meeter for this purpose, to call a *poynt or prycke*, that small printe of penne, pencyle, or other instrumente, whiche is not moued . . .' Recorde then marks three points on the page thus ∴ so that the reader can see for himself what they might look like. He admits that each of these dots has length and breadth 'but smal, and therfore not notable'. When he comes to a line, he does not even pay lip service to Euclid's definition of a breadthless length, but goes straight to a more intuitive description of a line as a series of points. Recorde even draws some of the points, ..., and invites the reader to fill in others to produce a continuous line, ————— . Only now does he remark that 'this *lyne*, is called of Geometricians, *Lengthe withoute breadth*', but observes that those who think so are engaged in 'mind workes' whereas the practical man needs to apply such things to 'handy workes'.

Recorde's definition of a straight line is simply 'the shortest that maye be drawenne betweene two prickes'. Once again the real teaching, however, is not in the definition but in Recorde's drawings:[6] the best part of a page, after a single very dull straight line, is filled with charmingly imaginative 'croked' lines, in the form of curves, spirals and wriggles. There is nothing like this in Euclid!

And so Recorde goes on, offering pictures and examples that relate his definitions to everyday objects: a die, a globe, an egg, a spire. At the same time he is continually changing Latin technical vocabulary into common English: 'a square angle', 'a sharpe angle', 'a blunt corner', 'a platte forme' (a surface), 'a croked platte', ' a bodie' and so on. His parallel lines are also called 'gemowe [twin] lines'; a tangent to a circle is a 'touche line'; opposite angles made by two crossing lines are 'matche corners', a sector removed from a circle is a 'nooke cantle', and so on in wonderful profusion.

After seventeen pages of such definitions we come to the practical part of the book, entitled 'The practike workinge of sondry conclusions Geometrical' (sigs C[.i]ᵛ–[I.ii]). This part consists of forty-six numbered 'conclusions', every one of which involves some kind of construction and so corresponds to Euclid's 'problems'. Conclusion 1 is precisely Euclid's Proposition I.1: 'To make a threlike [equilateral] triangle on any lyne measurable'. Like Euclid, Recorde gives a diagram and instructions. Indeed, his instructions are essentially the same as Euclid's but he presents them in much friendlier language: 'stay the one foot of the compas in one of the endes of that line, turning the other vp or down at your will, drawyng the arche of a circle'. Unlike Euclid, however, Recorde offers no proof that the triangle so constructed is equilateral. Instead he simply proceeds (in Conclusion 2) to give further instructions for a 'twileke' [isosceles] or 'noue-like' [scalene] triangle on any base.

Recorde's Conclusion 3 is Euclid's Proposition I.9: how to divide any angle *ABC* into two equal parts. Again, he begins with the same directions as Euclid: draw an arc with centre at *B*, crossing *BA* and *BC* at *D* and *E* respectively. But then he instructs that *DE* should be divided in half without actually explaining how this is to be done. More, and more serious, deviations from Euclid soon follow. Conclusion 5 explains how to draw a 'plumme line' [perpendicular] from a point *C* on a line *AB* (Euclid I.11). As before, Recorde at first gives the same instructions as Euclid, but then turns to the case where *C* may be so near the end of *AB* that the construction as he has described it is difficult or perhaps not possible. For this case he suggests an alternative, which depends essentially on constructing a 3–4–5 triangle with its right angle at *C*. This is a practical enough thing to do, but it is certainly not Euclidean. Nor are Recorde's instructions for drawing a plumb line using a set-square in Conclusion 9; or for constructing a plumb line from the centre of a bridge, in Conclusion 10, by finding the midpoint on the ground between the pillars of the bridge, and then dangling a 'long line with a plummet' (a weighted string) directly over it from above. These are not exposi-tions of Euclid but Recorde's attempt to relate Euclidean geometry to the practi-cal experience of his readers.

In the later part of the book, the practical gives way to the theoretical as Recorde works his way through most of the constructions given by Euclid: of squares and parallelograms equal to certain other figures, of tangents to circles and bisections of arcs, and of the various inscribed and circumscribed figures of Book IV of the *Elements*. Sometimes, however, one feels that Recorde implicitly draws on the reader's existing experience. When he comes to 'cinkangles' [pentagons], for example, his instructions become perfunctory, as though he expects the reader to know already what to do. The aim of Conclusion 37 is to construct an isosceles tri-angle in which the two base angles are each double the angle at the apex. The reader is told to divide the circumference of a circle into 'fyue equall partes' and then join division point 1 to points 3 and 4. How the five equal parts are to be determined, however, is not specified. Conclusion 38 is then concerned with constructing a regular pentagon in a circle, but Recorde tells us only that we should divide the circle into five equal parts, as in Conclusion 37. Clearly this is to be left to practical ingenuity. His final Conclusion (46) is also the final proposition (IV.16) of Book IV of the *Elements*: to construct a regular fifteen-sided polygon inside a circle, even though the problem of first dividing the circle into five has not yet been overcome.

Thus Recorde offers a blend of the theoretical and the practical, encouraging readers both to draw on their existing experience and to try out new constructions. It is easy to criticize his omissions and lapses, but before doing so one should recall once again what he was trying to do: to present a completely new subject to a completely new audience. We do not know who in the end read or used the first book of the *Pathway*, but if Recorde's readers really did comprise 'all sortes of men', some of them would surely have learned very much more than they knew before. Indeed, some would have had their first taste ever of classical learning.

The second book of the *Pathway* opens with another 'Preface' (sigs a.ij–[a.iiij]), in which Recorde explains that he wrote the first book some five years earlier (that is, around 1546) and had intended to follow it with applications of the techniques taught there. Now, however, he has decided to present the theorems on which his 'conclusions' were based. It has been suggested that Recorde's separation of problems from theorems was influenced by Proclus, whose commentary on the first four books of the *Elements* was published for the first time in the 1533 Grynaeus edition.[7] Proclus argued that although theorems are superior to problems, it is better to present problems first for those who are coming to the subject from a practical point of view. Recorde argued similarly. One might think, he claimed, that causes (the theorems) should go before effects (the 'conclusions', or constructions); but he could justify the reverse pedagogically: 'in order of teachyng the effect must be fyrst declared, and than the cause therof shewed, for so shal men best vnderstand things.' Recorde makes a convincing case but just at the moment where the reader is persuaded to agree with him, he adds a more homely reason for the order of his two books: he wrote the book of 'conclusions' five years ago, and only later decided what should follow it. Recorde was a master of pedagogy, but he was also thoroughly human.

. In the same Preface, Recorde also offered pedagogical reasons for omitting proofs and demonstrations from both books of the *Pathway*. Experience had shown him, he claimed, that assimilating both a proposition and its demonstration was 'a greate trouble and a painefull vexaction of mynde to the learner, to comprehend bothe those thinges at ones'. There was much to be said, therefore, for first learning to understand the sense of a proposition and only then grappling with its proof. Here he cited as authorities both Rheticus and Boethius, both of whom, he said, set forth whole books of Euclid without any demonstrations. Recorde himself was about to do the same, though if it later seemed worthwhile he would not refuse to provide 'sundry varietees of demonstrations, bothe pleasaunt and profitable also'.

The geometry of the second book begins not yet with theorems but with 'certaine grauntable requestes' (sigs b.i–b.ij). These are the 'postulates' of Book I of the *Elements*, starting with the first, the assertion that 'from any pricke to one other, there may be drawen a right line' and continuing to the fifth, the 'parallel postulate'. Recorde also has a sixth postulate, added by some later interpreters of Euclid, that a surface cannot be formed from two straight lines. Then he moved on to 'common sentences, manifest to sence, and acknowledged of all men', in other words, Euclid's 'common notions' (sigs b.ij^v–c.[i]). He presents all the common notions from Book I of the *Elements* but also some extra 'sentences' concerning inequality: 'When euen portions are added to vnequalle thinges, those that amounte shalbe vnequall.' The penultimate 'sentence' is Euclid's 'Euery whole thing is greater than any of his partes,' but the final one is not Euclidean, and not even true under the Euclidean definition of a 'part' as a unit fraction of the whole: 'Euery whole thinge is equall to all his partes taken togither.'

Finally, there follow seventy-seven 'theoremes of Geometry' (sigs c.[i]–m.iv), all of them propositions from Books I to IV of the *Elements*. The modern reader will find in the second book of the *Pathway* many familiar rules and theorems: rules for determining congruent triangles, for instance, or the theorem that triangles on the same base and between the same parallels are equal in area (theorem 27), or that an angle at the centre of a circle is twice the angle at the circumference standing on the same arc (theorem 64). Most of the theorems from Books I to IV of the *Elements* are included but not in the careful deductive order in which Euclid had placed them: sequencing no longer matters since there are no proofs. Nevertheless Recorde takes care to clarify the meaning of each theorem by means of a lettered diagram and an explanation of what the theorem says about it. The statement of 'Pythagoras' theorem', for instance, is accompanied by a diagram of a 3–4–5 triangle with the squares on its sides divided into grids of 9, 16 and 25 smaller squares, respectively. One only has to count and add the smaller squares to understand the meaning of the theorem. This may not be rigorous deductive geometry but it certainly fulfils Recorde's purpose of teaching his reader the 'sense' of the theorems. If the word 'demonstration' is interpreted not as a technical 'proof' but in the more informal sense of a 'practical explanation', then Recorde's examples are very often perfectly good demonstrations of his theorems.

Even a modern reader can still learn a good deal of elementary geometry from the *Pathway* and there is no reason to suppose that the same would not have been true for Recorde's contemporaries. Any of his 'carpenters, carvers, joiners, or masons' who could read would surely have been taken quite some way along the path to knowledge by Recorde's homely language, practical examples and easy explanations. Recorde presented none of the formal deductive structure of the *Elements* but he did offer the common English reader a first insight into the branch of mathematics called geometry, and indeed much of the content of the first four books of Euclid. He did not translate Euclid's *Elements* literally but he did translate them conceptually, for a readership very different from any that Euclid could have envisaged.

Later English editions of the *Elements*

The Pathway to Knowledg did not have the long-running success of Recorde's arithmetic textbook, *The Ground of Artes* (1543), but proved sufficiently popular for two further editions to be published after Recorde's death, in 1574 and 1602, both for the London bookseller John Harrison.

Recorde was the first writer to expound Euclid's *Elements* in English, but he was certainly not the last.[8] Twenty years after the publication of the *Pathway*, Henry Billingsley brought out his lavish and beautifully printed *The Elements of Geometrie of the most Auncient Philosopher Euclide* (1570). It is now perhaps best known for its preface by John Dee, classifying the various branches of mathematics in a great 'groundplat'. The text itself was prepared by Billingsley from the 1516 Campanus–Zamberti edition.[9] Billingsley was an astute scholar; at the same time he, like Recorde, was aware of the needs of those for whom this was new material. Between the propositions he included a great deal of explanatory commentary, and also provided charming 'pop-up' diagrams to give the reader a practical insight into some of the theorems of three-dimensional geometry.[10] Just as Recorde encouraged the reader to take up his pen and fill in the points on a line, Billingsley too wanted his readers to engage in 'handy workes' as well as 'mind workes'.

By the middle of the seventeenth century, English editions of Euclid began to proliferate, with Thomas Rudd's *Euclides Elements of Geometry* (1651) and Isaac Barrow's *Euclide's Elements* (1660). Barrow, writing for young Cambridge students, followed Euclid's structure but, as Recorde had done, wrote in homely language, with easy explanations of what he considered obscure or difficult matters. He also introduced some elementary algebraic symbolism, which for contemporary readers must have seemed to modernize the contents considerably. John Leeke and George Serle brought out their *Euclid's Elements of Geometry* in 1661, Reeve Williams and William Halifax each published translations of a French edition of the *Elements* by Dechales in 1685, and these were followed by William Alingham's *Epitome of Geometry* in 1695.

Further English editions followed throughout the eighteenth and nineteenth centuries. Each editor wrote in his own style, and with his own ideas on how best to present the *Elements* to new readers. Perhaps the edition that came closest in spirit to Recorde's was Oliver Byrne's *The First Six Books of the Elements of Euclid* (1847), a beautiful coloured edition in which the propositions are demonstrated with clever use of red, blue, yellow and black. As in the *Pathway*, the written proofs take second place to the diagrams, which attempt to make the sense of the propositions immediately visible to the eye. Byrne's edition was certainly the most user-friendly edition of the *Elements* since Recorde's.

The tradition of making Euclid accessible to as wide an audience as possible has continued into the twenty-first century with David Joyce's interactive online version. Like almost every earlier editor, Joyce provides extensive explanatory commentary, but now combined with some of the best ideas of Byrne and Billingsley: colour and moving parts. And like Recorde, Joyce encourages readers to engage with their hands, no longer with pen and ink but with touch pad or mouse.

Euclid's *Elements* are now easily available to anyone anywhere with a reasonable command of English and access to a computer. This is a very far cry from the situation before 1551, when the *Elements* could be read only by a tiny handful of educated readers. In England, the move towards greater accessibility was initiated by Recorde, who had to begin from the most basic level, first explaining what the strange new subject of geometry was about and why anyone should trouble to study it, and then teaching its foundational rules and constructions. Seen from this perspective, Recorde's *Pathway*, which at first may appear a somewhat crude and incomplete version of the *Elements*, instead comes to be recognized as a courageous and intelligent attempt to convey a hugely significant part of Classical learning to ordinary readers of English.

Notes

1   On the early printed editions of Euclid see Thomas Little Heath, *The Thirteen Books of Euclid's Elements*, 1908, vol. 1, pp. 91–113; Charles Thomas-Stanford, *Early Editions of Euclid's Elements* (London, 1926), reprinted with additional plates by Alan Wofsy Fine Arts (San Francisco, 1977).

2   See Jack Williams, *Robert Recorde: Tudor Polymath, Expositor and Practitioner of Computation* (London, 2011), ch. 8, pp. 106–8.

3   Today five copies of Zamberti's edition of Euclid are held in Oxford college libraries (Corpus Christi, Magdalen, Merton, New, St John's) and one in Cambridge (Trinity).

4   In this essay all definitions and propositions from the *Elements* are given in English from *The Thirteen Books of Euclid's Elements*, edited by Thomas Little Heath, which has remained the standard English translation. For a modern interactive version, however, the reader is encouraged to use and experiment with David Joyce's edition at *http://aleph0. clarku.edu/~djoyce/java/elements/trip.html*

5   This may be transcribed in more modern English as follows: 'Excuse me gentle reader if anything is amiss, strange paths are not trodden all truly at first: the way must necessarily be cumbersome, where none has gone before, where no man has given light, light is it to offend, but when the light is shown once, light is it to amend. If my light may so light some other, to spy and mark my faults, I wish it may so lighten them that they may avoid offence.'

6   For an excellent discussion of Recorde's use of diagrams see Michael J. Barany, 'Translating Euclid's diagrams into English, 1551–1571', in Albrecht Heeffer and Maarten Van Dyck (eds), *Philosophical Aspects of Symbolic Reasoning in Early Modern Mathematics* (London, 2010), pp. 125–63.

7   Francis R. Johnson and Sanford V. Larkey, 'Robert Recorde's mathematical teaching and the anti-Aristotelian movement', *Huntington Library Bulletin*, 7 (1935), 59–85.

8   For further information on all the English editions mentioned here and many others see June Barrow-Green, '"Much necessary for all sortes of men": 450 years of Euclid's *Elements* in English', *BSHM Bulletin*, 21/1 (2006), 2–25, the article that provided the original inspiration for this chapter.

9   See R. C. Archibald , 'The first translation of Euclid's *Elements* into English and its source', *American Mathematical Monthly*, 57 (1950), 443–52.

10   See Katie Taylor, 'Vernacular geometry: between the senses and reason', *BSHM Bulletin*, 26/3 (2011), 147–59.

FIVE

The Castle of Knowledge: astronomy and the sphere

STEPHEN JOHNSTON

Robert Recorde's *The Castle of Knowledge* was published in 1556, just a few years after the appearance of Nicolaus Copernicus' *De revolutionibus orbium caelestium* (1543). As one of the first books to comment publicly on the new heliocentric theory, Recorde's *Castle* has attracted regular attention since the nineteenth century. When significance in the history of science was judged only by an author's contribution to progress, the way to assess a sixteenth-century astronomy textbook was by its stance on Copernicanism. Thus, for the earliest historians who examined the *Castle*, the questions seemed simple and stark. What side of the fence was Recorde on? Was he for progress and modernity, or conservatism and tradition?

It proved surprisingly difficult to provide compelling answers. Even after a new standard was set for the study of Recorde by the literary historian Francis Johnson in the 1930s, unanimity did not emerge. Johnson concluded that Recorde was a convinced Copernican, albeit one who still gave pedagogic prominence to the older geocentric worldview.[1] But Recorde was then subsequently identified as an open-minded but unshakeable upholder of Ptolemaic tradition.[2] This lack of agreement was due not only to the difficulty of the evidence but also to the inadequacy of the question, which misses a significant part of Recorde's purpose in airing the new cosmological doctrine. Although only a brief passage in a larger work, and one that has often been discussed, it is worth starting with Recorde's mention of Copernicus. If this helps to set aside any anachronistic or inappropriate expectations of the *Castle*, the way will be left clear for a broader and more sympathetic treatment of the work.

Recorde on Copernicus

Copernicus appears relatively late in Recorde's volume, in the last of its four books. By this stage we have been not just introduced to, but thoroughly instructed in, the order and hierarchy of the traditional geocentric world. The very fact that Copernicus was seen to pose a challenge shows that Recorde was a discriminating reader. He presents Copernicus' doctrine as a claim about the physical world, rather than merely a mathematical fiction designed to provide predictions of the positions of stars and planets. Recorde was evidently unpersuaded by the anonymous foreword to *De revolutionibus* (inserted during printing by Andreas Osiander), which notoriously declared that astronomy was concerned only with the correct calculation of astronomical appearances rather than the true causes of the celestial motions.[3]

Couched in the familiar dialogue form he used elsewhere, Recorde subtly dramatized the encounter with heliocentrism through the interplay between the Master and his Scholar. By following the twists and turns of the text in detail we can get a first flavour of the style and tone of the *Castle*. When the Master reaches the topic of the immobility of the earth (p. 164), he thinks there is no need to spend any time proving it. Why labour to demonstrate what no one questions, particularly since the 'opinion is so firmelye fixed in most mennes headdes, that they accompt it mere madnes to bring the question in doubt'. It is the Scholar who enters a cautionary note: 'Yet sometime it chaunceth, that the opinion most generally receaued, is not moste true.' This opening gives the Master the opportunity to name a few of the ancients who indeed held the contrary opinion. But he does not want to enter into the arguments on either side. The Master's mind is evidently not so firmly fixed as that of most men, for he considers that the proponents of the earth's motion had a strong case. However, their 'reasons are to[o] difficulte for this firste Introduction, & therfore I wil omit them till an other time'. He likewise passes over the arguments for the earth's stability, though not without noting that these reasons 'doo not proceede so demonstrablye, but they may be answered fully, of him that holdeth the contrarye'.

The Master then hastily qualifies this apparent stalemate by remarking that it applies only to the possibility of the earth's rotation on its own axis at the centre of the world. Any motion which takes the earth away from the centre of the world would fall victim to the 'invincible reasons' – Ptolemy's observational arguments – which the Master had presented in the immediately preceding section. With those arguments for the earth's centrality fresh in mind, the Scholar immediately appreciates the distinction and recalls that 'if the earthe were alwayes oute of the centre of the worlde [e.g. if it had an orbit], those former absurdities woulde at all tymes appeare.' This is the point at which the Master introduces

Copernicus, who has not only made the earth rotate on its own centre but also stated that it 'may be, yea and is, continually out of the precise centre of the world 38 hundreth thousand miles' (p. 165). Without drawing out the implications of this huge orbit, the Master simply reiterates that the controversy requires more profound knowledge than can be presented in his introductory text, so he will let it pass until a later opportunity.

The Scholar, who had only a moment before been ready to challenge common sense and general opinion, cannot stomach the enormity of Copernicus' claim. 'Nay syr in good faith, I desire not to heare such vaine phantasies, so farre againste common reason, and repugnante to the consente of all the learned multitude of Wryters, and therefore lette it passe for euer, and a daye longer.' Now it is the Master who displays the higher wisdom. The Scholar is too young to be a good judge in such a matter and should not condemn what he does not fully understand. When the Master can return to expound Copernicus fully 'you shall not only wonder to hear it, but also peraduenture be as earnest then to credite it, as you are now to condemne it.'

There can be no doubt that Recorde was remarkably open-minded and sympathetic in his response to Copernicus, but he does not commit himself. To try to infer his own settled view (if he even had one) from this passage runs counter to the larger lesson he is trying to teach. His deferral of the issue is not hesitancy or timidity but a refusal to abuse his own authority. To assert either the truth or falsity of heliocentrism at this point in his programme of instruction would be to require assent without argument. The Scholar is inadequately prepared to understand let alone assess the merits of the arguments on either side. Only when he has acquired the necessary intellectual equipment can he reach a genuinely informed judgement. To pretend otherwise is to misunderstand not only the moral position of the teacher but the proper order and method of teaching itself.

Recorde could have won himself far greater posthumous fame (or notoriety) if he had written more extensively on Copernicus. That he did not do so was largely because of his concern with due method in teaching and his care not to overwhelm the student with matters that would need to be taken on trust. When the Master cautions the Scholar against hastily rejecting what he does not comprehend, Recorde probably wants the reader to recall the case of the eminent Church Father Lactantius, who served as Christian adviser to the emperor Constantine in the early fourth century. The *Castle* gives much space to discussing and rebutting Lactantius' mocking rejection of the sphericity of the earth and the heavens, which Copernicus had also sharply criticized. Whatever his spiritual authority, Lactantius was uninstructed in astronomy and philosophy and guilty of the most childish of errors. We therefore need to exercise care with all writers, whether authoritative in religion or even in astronomy:

No man can worthely praise Ptolemye, his trauell being so great, his diligence so exacte in obseruations, and conference with all nations, and all ages, and his reasonable examination of all opinions, with demonstrable confirmation of his owne assertion, yet muste you and all men take heed, that both in him and in al mennes workes, you be not abused by their autoritye, but euermore attend to their reasons, and examine them well, euer regarding more what is saide, and how it is proued, then who saieth it: for autoritie often times deceaueth many menne. (p. 127 [129])

Sacrobosco and the sphere

Recorde must be judged as a teacher rather than a modern scientist in pursuit of progress. If we are to respect the integrity of his work we cannot restrict all our attention to an admittedly fascinating but nevertheless isolated passage. What was *The Castle of Knowledge* actually about, and what was Recorde's genuine contribution in writing it? Recorde himself points the way on the reverse of the title page where he gives an overview of the volume, describing it as devoted to 'the explication of the sphere bothe celestiall and materiall'. His terms may no longer be familiar – 'the sphere' does not appear in the modern school curriculum – but it was nevertheless one of the most popular medieval and Renaissance 'scientific' topics, and one of the few mathematical subjects to which undergraduates were regularly exposed. To understand Recorde's text means understanding something of this long and rich tradition.

The name indissolubly linked to the literature of the sphere is that of Johannes de Sacrobosco. Although not the first to write on the subject, his *De sphaera* became the standard Latin textbook. Written in Paris in the early thirteenth century, it survives in literally hundreds of medieval manuscript copies. It was the first significant work on astronomy to be printed, appearing in 1472, and was issued in innumerable editions all over Europe for the next two centuries.[4] Already in the thirteenth century it was the object of commentary, and this grew into a veritable industry. By Recorde's time, the disproportion between original and accompanying text was becoming almost comical. Sacrobosco's work is only about 9,000 words long – a modern-day academic article. The leading Jesuit mathematician Christopher Clavius brought out a commentary in the later sixteenth century, which he repeatedly enlarged and reissued and which eventually weighed in at over 800 pages.

Sacrobosco's *De sphaera* provides a brief introduction to its subject, in four chapters.[5] The first opens with the geometrical definition of a sphere and the division of the universe into the heavens and the sublunary realm of the four elements. It provides arguments for the spherical form of the heavens and the earth and outlines the number and motions of the celestial spheres of the planets and stars. The earth is the immobile centre of this vast machine, and a mere point in relation to the size of the firmament. Chapter 2 focuses on the circles of the celestial sphere: the celestial equator or equinoctial, the ecliptic with its zodiac signs, the tropics, colures and arctic and antarctic circles, as well as the horizon and meridian which are particular to a given location on earth. The third chapter deals with the various risings and settings, garnished with liberal quotations from the classical poets. It also tackles the variation in the length of the day, both throughout the year and according to the elevation of the pole (latitude). This latter variation provides a quantitative way to divide the habitable part of the earth into seven 'climates', bands stretching east to west on the earth. The climates are defined so that the longest day on the centre line of each climate is half an hour longer than that of the equivalent line on the preceding climate. The final chapter provides a very compressed outline of solar and planetary motions, introducing the apparatus of equant, deferent and epicycle and the notion of the retrograde motion of the planets. The work ends with a brief explanation of eclipses and a concluding note that the solar eclipse marking Christ's crucifixion must, on astronomical grounds, have been a miraculous event.

Although much the most popular author on the sphere, Sacrobosco was not the only one. Each author and commentator offered a slightly different vision of the subject, with a different range of topics and emphases. Moreover, the genre was supplemented by Renaissance works on cosmography, a subject which repackaged much the same material under a new and fashionable title and similarly tied together the heavens and the earth in its presentation.

The *Castle* and the sphere

Recorde used the longevity and diversity of the tradition of the sphere as the starting point for his dialogue. The Scholar reports to the Master that he has been reading various works about the world and its parts. Rather than illumination, the experience has filled him with confusion and doubts. He tried Proclus on the sphere, which begins by defining the axis of the world, without saying what the world is. Dispirited at such a disorderly beginning he turned to Sacrobosco, who opens with the definition of a sphere, 'but nothinge lyke to that sphere, whiche I before had bought, as an apt instrument to learne by' (p. 3). The Scholar had presumably purchased an armillary sphere, quite different in appearance from the solid sphere defined by Sacrobosco. So he moved on to the cosmography of Oronce Fine (Orontius) – to no avail: it did 'disagree from them bothe: and generallie, euerye one from other, so that I know not wher to beginne'.

In its vivid details and empathy with the predicament of the beginner, Recorde's characteristic style is evident from the outset. But he also artfully invokes the full range of the tradition to which he is responding, from the ancient Greek text of Proclus, through the classic medieval textbook of Sacrobosco and on to the contemporary treatment by the French royal professor of mathematics, Oronce Fine.[6] The disorientation of his student provides the rationale for a new contribution to this rich literature.

Here, as elsewhere, Recorde considered that one of his main responsibilities was the proper arrangement of his material. 'In ordre of teaching is more credit to be gyuen to a master, then in affirming of anye doctrine: for the ordre is by longe experience best knowen of such men' (pp. 100–1). We have no independent evidence for Recorde's teaching but his books leave no doubt that it was extensive and carefully considered. Just as in his geometry, he planned the *Castle* as a deliberate sequence, beginning with concepts and results and leaving details and proofs until the later stages.[7] Although the volume is divided into four books, these are not equally proportioned: Books I to III are dwarfed by the last book, which occupies two-thirds of the whole and which promises 'the proofes of all that is taught before, and other diuers notable conclusions annexed therto. but nothing in a manner with out demonstration and good proofe' (p. 97). This lengthy display of 'demonstration and good proofe' is not Euclidean in style but consists of informal arguments and numerical tables which give the full quantitative detail that would have overwhelmed the earlier general enunciations.

The first of the four books provides the 'big picture' of the universe and its parts, without entering into any doubtful matters. We are led through the sequence of the stars and planets and told of the central immobility of the earth as the realm of the four elements. But whereas Sacrobosco had mentioned a sphere beyond that of the fixed stars to account for the precession of the equinoxes, Recorde deliberately leaves such advanced matters for later treatment. He then presents the principal points and circles of the celestial sphere.

With these traditional doctrines in place, Book II switches to a much more practical register by discussing the making of two forms of sphere. The first is a solid globe, the second an armillary sphere with all the circles specified in the previous book. For the globe, Recorde offers practical tips on turning and testing the sphere, with details such as the form of the stand and the addition of clamps for the meridian ring, as well as instructions and cautions on the laying out of the lines. This level of material detail was not a core component of the tradition of the sphere: Sacrobosco barely mentions material spheres, though Oronce Fine did include such constructional matters in his discussions of practical geometry.[8] Although fuller than most treatments, Recorde nevertheless acknowledges that 'manye thynges in the makinge, and in the vse also of instrumentes, are better

perceaued by a lyttle sighte, then by many woordes' (p. 54 [34]). The armillary
sphere is presented more briefly, with only a schematic diagram (Figure 1 shows
a labelled example of such a sphere from a later text). To ease construction,
Recorde provides a woodcut showing all the circles drawn to the correct scale
and divided into their degrees. A beginner could simply copy or rescale these to
minimize labour and to preserve the correct proportion between all the parts.

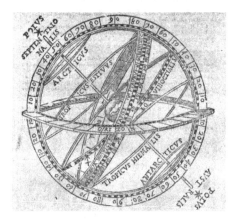

Figure 1. An armillary sphere; from the third edition of
Christopher Clavius' *In sphaeram Iohannis de Sacro Bosco commentarius* (Rome, 1581).
Whipple Library, University of Cambridge

Book III is on the basic uses of the sphere, showing how the earth has its own
circles and zones equivalent to the celestial ones above. The Scholar finds this
easy, as long as the material sphere is to hand, vindicating the Master's order of
instruction: 'For that cause did I teach you the making of it, before I instructed
you in the vse of it' (p. 66). As in Sacrobosco's third chapter, Recorde here also
discusses the determination of latitude from the height of the pole (or another
star, if necessary) and the diversity of days in different regions and times of
the year. Finally, there is Book IV, which offers a fuller and much more detailed
presentation of all that has gone before. There are the promised arguments for
the roundness of the heavens and the earth, and for the centrality, immobility and
immeasurably small size of the earth in comparison with the heavens. Recorde
discusses problems such as the differing definitions of the arctic and antarctic
circles given by the ancient Greeks and the medieval Latins, and argues against the
ancient opinion that the earth has uninhabitable zones. The co-ordinate systems
of positional astronomy – longitude and latitude, declination and place in the
zodiac – are introduced along with tables for the climates and lengths of day.

Whereas Oronce Fine had dispersed numerical tables throughout his *De mundi sphaera*, Recorde very deliberately restricted them to this last book, along with worked numerical examples to check the Scholar's understanding. Among the other topics covered are astronomical and poetical ascensions; equal and unequal hours; natural and artificial days; the sun's rising and setting; and the constellations and their stars. In a self-conscious mirroring of Sacrobosco, Recorde ends with a nod towards the motions of the planets and an introduction to eclipses, though confessing that Sacrobosco's 'woordes are shorte and therefore obscure, and so should my wordes be. [B]eside that, it is a disordrely forme to put the carte before the horse' (p. 279). Although ideally these matters would be deferred for later treatment, here he bows to precedent in including a brief discussion.

Pedagogy and scholarship

From the bare recital of his topics and their close correspondence with Sacrobosco's, it is clear that Recorde's achievement was a fresh arrangement of a well-defined body of traditional doctrine. The author's skill lay in grasping and removing the student's obstacles to understanding, and the *Castle* is deeply imbued with Recorde's pedagogic vision. This emerges in such features as the dialogue format, the interplay between Scholar and Master being lively and even sometimes entertaining, rather than in the stilted and forced style to which the genre can on occasion descend.

To make his material memorable and accessible Recorde plays to his English audience in his choice of geographical examples. Thus, in an argument from the ancient Greek author Cleomedes, he notes the distance between two towns, one in Greece and the other in Egypt. The Scholar follows the line of reasoning but thinks that 'I shuld better haue conceaued it, if I had knowen the two places whiche hee alleageth for examples sake' (p. 121). So the Master reworks the example with Newcastle upon Tyne and Arundel Castle, saying that he had once taken note of the latitude of the latter in riding that way. Geographical material could also provide immediacy and topical relevance. English traders to Guinea went to latitudes where the direction of shadows varies throughout the year (north in winter, south in summer), prompting the Scholar to muse: 'I heare saye, by our owne cuntrye men, whiche trauaile to Guinea, that they wente beyonde the sonne, whiche alwaies I tooke to be a lye of libertye permitted to farre trauelers, but now I perceaue it maye be true in one sense' (p. 86). Likewise, in discussing the differing lengths of the day, Recorde takes Wardhouse (Vardø, in the extreme north-east of Norway) as an example, 'where our newe venterers into Moscouia do touch in theyr viage'. At that latitude there is continual summer daylight for 73 days, at which news the Scholar exclaims 'This is meru-ailous straunge to me' (p. 75).

Recorde's eminently approachable text was geared towards easier apprehen-sion and retention of arguments and details. Twice in Book I and at the ends of Books III and IV, the Scholar repeats the main points that he has learnt, so that the message is not lost in the pleasurable dialogue form. (Recorde says that the practical Book II needs no such repetition.) He also kept his subject matter within tight bounds, carefully policing the boundaries between differ-ent topics. Although Oronce Fine titled his work 'The Sphere of the World, or Cosmography', Recorde kept his treatment of the sphere distinct from what he considered the separate grouping of cosmography and geography. He therefore repeatedly deferred specific topics to a future cosmographical treatise (e.g. pp. 100, 154, 175, 193, 272 [572]). In the same vein, when there is a risk of stray-ing too far into the realm of planetary motion, the Scholar is told he must wait until the 'theorics' of the planets are treated (pp. 60, 278 [286]). But because Recorde was following his own didactic sequence, he can borrow conclusions that he had already established elsewhere, particularly geometrical results from *The Pathway to Knowledg* (e.g. p. 120). Although his *The Gate of Knowledge* was never published, he took from it the description of the simplest form of quadrant for taking the height of the sun and moon (p. 68). Indeed, it was at the beginning of the *Castle* that Recorde published his poem 'An admonition for the ordrely trade of studye in the Authors woorkes, appertainyng to the mathematicalles', which specifies the sequence in which his books should be read.

However Recorde did not think that his own writings contained all the answers. Throughout the *Castle* he acted as a veritable reader's guide to the lit-erature of the sphere – comparing, explaining and correcting statements from a whole host of writers. Unlike many Renaissance authors, who deliberately concealed their sources, Recorde openly displays his reading, including schol-ars now extremely obscure within the history of science.[9] Rather than a proud boast of his own learning, this is presented as an act of charity. He hopes that others can thereby avoid his own experience: 'as the numbre of writers are infinite, so haue I founde great tedious payne in readinge a greate multitude of them' (p. 98). Recorde gave constructive but critical advice on further reading, suggesting the best works to consult after Proclus, Sacrobosco and Orontius and only hesitating from recommending a range of medieval English authors because their works were not yet printed. The Master also offered some cau-tions: 'As for Plinye, Hyginius, Aratus, and a greate manye other, [they] are to bee readde onlye of masters in suche arte, that can iudge the chaffe from the corne. and Ptolemye that worthye writer and myracle in nature, is to[o] harde for younge schollars' (p. 99).

All this might suggest an image of Recorde as a clear-sighted but friendly teacher. Yet he was more too. The *Castle* shows that he did not sacrifice scholarship for the sake of accessibility. Over and over again he takes up textual questions in the ancient sources in just the way that his professional colleagues in the College of Physicians were doing for Galen. Indeed, he had before him the English example of the distinguished medical humanist Thomas Linacre, whose first published work was a Latin translation of Proclus on the sphere, issued in Venice in 1499 as part of a collection of Greek astronomical writings. As Proclus was one of the authors who caused the Scholar's initial bewilderment, his text is repeatedly cited by Recorde, often in the Greek, followed by the Latin of 'our worthye contrye man D. Linaker' (p. 20), before Recorde's own English version. Nor was Recorde simply acting as a translator from Latin to English. He also demonstrated his ability to engage with and emend the original Greek behind Linacre's translation. In the discussion of constellations he noted 'a place in Proclus very much corrupted, whiche nowe I will only correct as I thinke good: and an other time will intreate more largely of it' (p. 269). When the Scholar remarks that Recorde's restoration is contrary to the common translation, the Master replies, 'and that common translation is as contrary to common sense.' He provided similar service in Greek for Cleomedes (p. 116) and at one point corrected the Latin translation of Strabo, though the problem here was due more to inadequacies of translation than to corruption of the Greek original (pp. 178 [171]–9). As ever, Recorde set such textual concerns in a larger frame. When the Master observed that Stöffler, Schöner, Copernicus, Reinhold and other major Renaissance astronomers were all deceived on a particular point by the old Latin translation of Ptolemy, the Scholar sagely picks up the Master's earlier theme of authority and reason: 'I thinke it (as manye thinges els be) is receaued by credite of authoritie, without disquisition of reason, whiche blyndeth manye wittye men oftentymes' (p. 270).

The moral value of astronomy

Recorde's ability to mix homely wisdom and moral precept with technical exposition is a striking feature of the *Castle*. The title page (see Figure 2) – much the most iconographically elaborate of all his published works – introduces themes that will recur in the dialogue.

Under the title is the castle, a high medieval round tower with Urania enthroned above two astronomers on the battlements, one with a quadrant, the other an astrolabe. The walls carry a framed but otherwise undistinguished example of Recorde's doggerel poetry. The real action flanks this central scene

Figure 2. Title page of a copy of *The Castle of Knowledge* owned by William Cecil.
Beinecke Rare Book and Manuscript Library, Yale University, Taylor 120

in the contrast between the personified figures of Destiny on the left and Fortune on the right. Destiny, governed by knowledge, stands solidly on a cube holding upturned geometric compasses in one hand, and an armillary sphere in the other. Fortune is blindfolded, ruled by ignorance and perpetually at risk of unbalancing on an unstable ball. She holds a slender thread attached to the handle of the wheel of Fortune, whose Latin motto '*Qui modo scandit corruet statim*' might be colloquially rendered as 'what goes up must come down' and suggests how Fortune will quickly lay low those who rise in the world. Between the two figures is a more pointed verse, which celebrates the triumph of celestial knowledge over earthly concerns:

> Though spitefull Fortune turned her wheele
> > To staye the Sphere of Vranye,
> Yet dooth this Sphere resist that wheele,
> > And fleeyth all fortunes villanye.
> Though earthe do honour Fortunes balle,
> > And bytells blynde[10] hyr wheele aduaunce,
> The heauens to fortune are not thralle,
> > These Spheres surmount al fortunes chance.

The conflict between Destiny and Fortune is not simply a title-page conceit, immediately forgotten once we enter the main text. A passage on Fortune and her rolling globe, as well as the contrast between Fortune and Destiny, is so extended that the Master exclaims 'I forget our purposed intent, with so many digressions of other bye matters' before resuming his thread (p. 114).

The moral power of astronomy is also set out in the Preface to the reader, the traditional place for an author's defence of his art. Recorde chooses to begin with the famous dream of Scipio, as recounted by Cicero, in which the grandson of the conqueror of Carthage returns to the scene of the earlier triumph. In a dream he ascends to the starry heavens where his ancestor is now deified and is shown the cosmos spread before him. From this celestial perspective the earth is unprepossessing and the great Roman empire pales into insignificance. With this classical sanction, Recorde makes astronomy the viewpoint from which earthly life should be judged: whoever will 'auoide the name of vanitie, and wishe to attayne the name of a man, lette him contemne those trifelinge triumphes, and little esteeme that little lumpe of claye: but rather looke vpwarde to the heauens, as nature hath taught him, and not like a beaste go poringe on the grounde' (sig. a.iiij).

The figure of Fortune on her ball and Scipio's dream were familiar tropes. Recorde had a tactical motive in making them emblematic of astronomy. Theologians and moralists were often tempted to condemn astronomy as a distraction from true spiritual preoccupations, treating it as a vain and impious prying into the secrets of creation. Recorde parries such judgements by proclaiming astronomy as a vehicle for true morality as well as the revelation of the visible heavens. It should properly be seen as a genuine incitement to religion: 'to studye to vnderstande [the heavens] ... but most of all to honour, praise and glorifie the author of them' (sig. a.v). Recorde also rehearses more prosaic reasons for asserting the value of astronomy: its utility for husbandry, navigation, medicine, time telling and the determination of Easter, not to mention more briefly specified uses in law, grammar, logic, rhetoric and history. Such arguments had a reassuring familiarity rather than bold novelty but were important in asserting

Figure 3. Plasterwork renderings of Destiny and Fortune at either end
of the Long Gallery at Little Moreton Hall, Cheshire, based on
the iconography of the title page of *The Castle of Knowledge*.
©NTPL/Andreas von Einsiedel

the value of the subject not just for specialists but for all citizens. The image of
the sphere, particularly in the form of the armillary sphere, took on an increasingly exemplary significance in the period as a symbol of knowledge and virtue.
It was used as a personal badge at the highest levels of society, already by Anne
Boleyn and much more so at the court of Elizabeth I, who was herself incarnated
as Urania.[11] While the increased popularity of the image of the armillary can
hardly be attributed exclusively to Recorde, there is one striking echo of the *Castle*
which does suggest the visual impact of its title page. The figures of Destiny
and Fortune were directly copied in the plasterwork of a new gallery added to
Little Moreton Hall, Cheshire about 1570–80.[12] This application of a moral-
ized astronomy as interior decoration (see Figure 3) suggests Recorde's persuasive
success in presenting the sphere as a guide to life as well as the world.

The *Castle* in English context

Recorde drew on an ancient, medieval and contemporary literature, and had
a European range of reference. But by publishing the *Castle* in the vernacular
he ensured his work would have only a local impact: English was very much
a marginal language in sixteenth-century Europe. Nevertheless, even if local,
that impact must have been considerable. The *Castle* was not only a cut above
Recorde's other publications but far outstretched anything on its subject previously produced by the English book trade.

As Recorde's only folio publication, the *Castle* was immediately distinctive and distinguished – in its physical size, its striking title page and many woodcut diagrams, its elegant Roman and italic fonts,[13] and its liberal use of Greek type for quotations from classical authors. Figure 4 shows the historiated letter T at the beginning of book I which, since it differs in size from the others in the book and shows an armillary sphere in use, was presumably cut specially rather than simply taken from stock. The work was clearly a prestigious production, and all the more remarkable given the difficult circumstances of Mary's reign for both Recorde and his habitual printer Reginald (Reyner) Wolfe, who was not only a close personal friend but shared Recorde's antiquarian and scholarly interests.[14]

The physical and visual distinction of the *Castle* would have been deeply impressive for those familiar with the London book market. Linacre's Latin translation of Proclus on the sphere had been reprinted in London in the early 1520s and an English translation by the notable Welsh humanist William Salesbury had appeared in 1550, *The Descripcion of the Sphere or the Frame of the Worlde*.[15] This was little more than a pamphlet: a work of some forty small pages in black letter and without diagrams. The printer obviously felt the lack of visual appeal, because he reissued the work a few years later with some woodcuts, at least two of which were geometrical diagrams for surveying and of no relevance to the text! If this suggests the low status of the project, Salesbury reveals just how unknown was its subject matter. At the start of the book he explains how he had come to make the translation, while also protesting his unfitness for the task – Welsh rather than English was his native tongue. He had been asked by a cousin for a treatise on the sphere in English, so he headed off to do some shopping at the centre

Figure 4. Letter T from *The Castle of Knowledge* (p. 1).
Beinecke Rare Book and Manuscript Library, Yale University, Taylor 120

of the London book trade around St Paul's cathedral: 'I walked myself rounde aboute all Poules churche yarde, from shop to shop enqueryng of suche a treactyse neyther coulde I here of any that eyther wrote of this matier proposely, nor yet occasionaly.' Determined to find something, he cast his net wider:

> by my fayth syr, I returned backe euen the same way (but wondrynge moche at the happe) and asked agayne for the same workes in latyn, whereof there were. iij. or foure of sondry Aucthors brought, and shewed vnto me, amonge all which (for the breuyte and playnes) I chose Proclus his doynge.[16]

Against such an impoverished backdrop, Recorde's *Castle* must have appeared an extraordinary achievement.[17] Its ambition presumably discomfited the physician and almanac maker Anthony Askham (Ascham), who had promised a treatise on cosmography in the 1555 edition of his annual almanac, which never appeared.[18]

Recorde also set the standard for future publications. As we saw, he envisaged his own treatise on geography and cosmography, but the only further text he issued was *The Whetstone of Witte*. It is difficult not to read William Cuningham's *The Cosmographical Glasse* of 1559 as a deliberate attempt to plug the gap left by Recorde's death the previous year. Another handsome folio by a mathematically-inclined physician, Cuningham's work was composed as a dialogue (though the Scholar and Master are dignified as Spoudaeus and Philonicus). He refers not only to Recorde's other mathematical books, but specifically to the way that the *Castle* includes almost all the arguments of earlier authors.[19] Like Recorde he quotes Cleomedes, Proclus and Theodosius in Greek, with both Latin and English translations, and he even mimics Recorde's technique of ending the individual books of the work with a repetition of the matters treated so far.

The impressive scale and presumably high cost of the *Castle* militated against it ever becoming a truly popular work, though it was reprinted in 1596. The little evidence that we have of its owners nevertheless suggests that it was able to engage a remarkably wide spectrum of readers, from seamen to ministers of state.

When Martin Frobisher set off on his elaborately equipped first voyage of 1576 in search of the North-West Passage, his working library included copies of both the *Castle* and Cuningham's *The Cosmographical Glasse*, which together cost the large sum of ten shillings.[20] This was a major expedition, whose list of expensive mathematical equipment suggests that the purchases were as much to impress the investors as for practical use at sea. That the *Castle* nevertheless did have something to offer English navigators and adventurers is suggested by a much later (and far more obscure) reference from the contested Caribbean. In 1633 the Spanish governor of the island of Margarita heard reports of an English settlement on the north-east tip of nearby Trinidad. He sent a military expedition to dislodge the interlopers: three companies of Spanish infantry and fifty native Indian archers destroyed the upstart English settlement, burning its fortifications and returning with prisoners, arms and other booty. Among the prizes triumphantly carried back to Margarita was a haul of seventy-seven books. The majority of the texts seized were suspected of harbouring heretical doctrines and were shipped off to the Inquisition – we know nothing of their identity or ultimate fate. The remainder were sent to Spain where, after being listed, a few deemed particularly useful were passed on to appropriate recipients. Amongst a number of now unidentifiable books on arithmetic and navigation the fifth on the list was 'Another entitled The Castle of Knowledge, which appears to be a treatise on the sphere, Vellum binding.'[21] Himself an early adviser to the Muscovy Company, Recorde would surely have been cheered to know that, nearly eighty years after its first publication, his text was still going to sea with English mariners.

It is hard to believe that these Caribbean-bound owners would have had much use for Recorde's emendations of corrupt passages in ancient Greek texts. But the book certainly had such readers. The copy reproduced in Figure 2 belonged to the statesman Sir William Cecil, whose slightly mutilated signature appears at the foot of the title page. The rapidly rising Cecil stood down from public business during the reign of Mary but under Elizabeth became firstly Principal Secretary and then Lord Treasurer. Cecil had been trained in the humanist Cambridge of the 1530s and retained his intellectual interests. He was warmly described in the preface to Roger Ascham's *The Scholemaster* (1570): 'Though his head be never so full of most weightie affaires of the Realme, yet, at dinner time he doth seeme to lay them always aside: and ever findeth fitte occasion to taulke pleasantlie of other matters, but most gladly of some matter of learning: wherein he will curteslie hear the mind of the meanest at his table.'[22] Cecil's vision of learning was broad and included technical subjects. In 1559 he sought the assistance of the English ambassador in Paris to procure books, noting: 'I am now and then occupied in Vitruvius de Architectura; and therefore if there be any writers besides Vitruvius, Leo Baptista [Alberti] and Albert Durer (all which three I have) I would gladly have them.'[23] He was sufficiently engaged with Recorde's *Castle* to annotate lightly both Books I and III, omitting the practical Book II and the detail of Book IV. The double spread in Figure 5 captures his style of annotation. The note on the left-hand page (partly lost through subsequent trimming) picks out the three authors foregrounded by Recorde (Proclus, Sacrobosco and Orontius) while the lack of consensus which so troubled the Scholar catches Cecil's eye at the top of the right-hand page: 'Disagreement of authors'. Finally, he pulls out the first Greek word from a quotation from Aristotle at the foot of the page: κόσμος (cosmos).

Figure 5. Annotations by William Cecil in *The Castle of Knowledge* (pp. 2–3).
Beinecke Rare Book and Manuscript Library, Yale University, Taylor 120

The combination of Caribbean evidence and the fortunate survival of Cecil's copy helps to avoid a reductive interpretation of the *Castle*'s character and fate. Certainly, as a vernacular text with much attention to material construction and numerical problem-solving, it could be used by those with practical and perhaps even urgent needs. But it also rested on scholarship that could be appreciated by an erudite administrator such as Cecil. The diversity of Recorde's audience and its appeal to an ideal of popular learning mirrored the new relationships being negotiated between knowledge and commerce, and between writers and artisans, as the economic and intellectual implications of the mechanical printing press were worked out.

Conclusion

Robert Recorde could easily have been forgotten. He cannot fill the stage in the role of revolutionary hero of science. Moreover, far from representing the triumph of virtue, his career and indeed life ended in personal disaster in the King's Bench prison in Southwark. Recorde alludes to the gathering storm at the end of the *Castle*: 'there was neuer any good Astronomer, that denyed the

Maiestie and prouidence of God, though many other denyed bothe: but nowe farewell for a time: I am dryuen to omytte teachinge of Astrononye, and must of force go learne some lawe' (p. 284). Recorde has the Master allude here to his own legal entanglements, which saw him clash with William Herbert, Earl of Pembroke.[24] This would become a matter of deadly seriousness rather than merely the play of words in a literary dialogue. Recorde was contending not with the regularity and order of the celestial realm and the sphere of Destiny, but living through turbulent times all too evidently governed by the wheel of Fortune.

In the Preface he had written that great downfalls would be advertised by signs and tokens in the heavens, which anyone could read. He was there speaking of such cataclysms as Noah's Flood, where the celestial signs revealed God's clear intentions, so that even those whom Noah had been unable to reach in his preaching had no excuse for inaction.

> So was there neuer anye greate chaunge in the worlde, nother translations of Imperies, nother scarse anye falle of famous princes, no dearthe and penurye, no death and mortalitie, but GOD by the signes of heauen did premonishe men therof, to repent and beware betyme, if they had any grace. (sig. a.v)

The irony is that, for all his astronomical acumen, Recorde did not foresee the unfolding of his own ruin. If he did not fully grasp the gravity of his predicament in the face of socially powerful adversaries, it was perhaps because his priorities were those of a remarkable and committed teacher who truly had faith that reason and knowledge would triumph over authority and ignorance. The vicissitudes of public life in mid-Tudor England more than tested such faith.

Notes

1   Francis R. Johnson, *Astronomical Thought in Renaissance England: A Study of the English Scientific Writings from 1500 to 1645* (Baltimore, 1937), ch. 5.

2   Louise Diehl Patterson, 'Recorde's cosmography, 1556', *Isis*, 42/3 (1951), 208–18. Recorde is also given even-handed treatment in the textbook account of Marie Boas (Hall), *The Scientific Renaissance 1450–1630* (New York, 1962; republished 1994), pp. 94–5. She comments that Recorde was no different from physics teachers of the early twentieth century, who continued to present Newton before turning to Einstein.

3   A now classic work on the complex relationship between mathematical astronomy and natural philosophy in the sixteenth century is Nicholas Jardine, *The Birth of History and Philosophy of Science* (Cambridge, 1984).

4   Olaf Pedersen, 'In quest of Sacrobosco', *Journal for the History of Astronomy*, 16 (1985), 175–220.

5   Lynn Thorndike, *The Sphere of Sacrobosco and its Commentators* (Chicago, 1949).

6   Alexander Marr (ed.), *The Worlds of Oronce Fine: Mathematics, Instruments and Print in Renaissance France* (Donington, 2009), particularly Adam Mosley, 'Early modern cosmography: Fine's *Sphaera mundi* in content and context', pp. 114–36.

7   The classic and still valuable account is Francis R. Johnson and Sanford V. Larkey, 'Robert Recorde's mathematical teaching and the anti-Aristotelian movement', *Huntington Library Bulletin*, 7 (April 1935), 59–87. See also chapter 4.

8   Pascal Brioist, 'Oronce Fine's practical geometry', in Marr, *The Worlds of Oronce Fine*, pp. 52–63; see pp. 60–1 for Fine's material details of the geometrical square.

9   For example his reference to Joachim Sterck van Ringelberg (p. 9), on which see Adam Mosley, 'Objects of knowledge: mathematics and models in sixteenth-century cosmology and astronomy', in S. Kusukawa and I. Maclean (eds), *Transmitting Knowledge: Words, Images and Instruments in Early Modern Europe* (Oxford, 2006), pp. 193–216, pp. 212–13.

10   These 'bytells blynde' echo the early modern phrase 'as blind as a beetle' and mirror the blindfolding of Fortune.

11   Jean Wilson, 'Queen Elizabeth I as Urania', *Journal of the Warburg and Courtauld Institutes*, 69 (2006), 151–73.

12   N. Pevsner and E. Hubbard, *The Buildings of England: Cheshire* (Harmondsworth, 1971), p. 257. My thanks to Jacqueline Stedall for first pointing out to me this plasterwork copy. Images of the remarkable half-timbered building are now most easily seen online, for example at *http://en.wikipedia.org/wiki/Little_Moreton_Hall* (accessed 11.02.11).

13   Curiously, apart from the title, the work's only significant piece of text typeset in traditional black letter is the letter of dedication to Queen Mary.

14   Andrew Pettegree, 'Wolfe, Reyner (d. in or before 1574)', *Oxford Dictionary of National Biography* (Oxford, 2004), online edn accessed 20.02.11.

15   R. Brinley Jones, 'Salesbury, William (b. before 1520, d. *c.*1580)', *Oxford Dictionary of National Biography* (Oxford, 2004), online edn accessed 20.02.11.

16   William Salesbury, *The Descripcion of the Sphere or the Frame of the Worlde* (London, 1550), sigs A.j^v–A.ij.

17   That there was a more widespread polite interest in the sphere by the early 1550s is suggested by a surviving manuscript English translation of Sacrobosco. This was made in about 1551 by William Thomas, another scholar with Welsh connections, for the teenaged second Duke of Suffolk; see Johnson, *Astronomical Thought in Renaissance England*, p. 133 and Dakota L. Hamilton, 'Thomas, William (d. 1554)', *Oxford Dictionary of National Biography* (Oxford, 2004), online edn accessed 20.02.11.

18   Bernard Capp, 'Askham, Anthony (*c*.1517–1559)', *Oxford Dictionary of National Biography* (Oxford, 2004), online edn accessed 20.02.11.

19   William Cuningham, *The Cosmographical Glasse* (London, 1559), pp. 49–51.

20   Richard Collinson, *The Three Voyages of Martin Frobisher*, Hakluyt Society (London, 1867), p. x.

21   All details in this paragraph are taken from Eleanor B. Adams, 'An English library at Trinidad, 1633', *The Americas*, 12/1 (1955), 25–41; the listing of Recorde appears at p. 38.

22   Cited by Graham Parry, 'Patronage and the printing of learned works for the author', in John Barnard and D. F. McKenzie (eds), *The Cambridge History of the Book in Britain*, vol. IV: *1557–1695* (Cambridge, 2002), pp. 174–88, p. 175.

23   Cited by Lynn White, Jr, 'Jacopo Aconcio as an engineer', *American Historical Review*, 72/2 (1967), 425–44, p. 430.

24   See chapter 1. Recorde's suit and Pembroke's counter-suit occurred at the same time as he was writing and publishing the *Castle*; see the dating established by Patterson, 'Recorde's cosmography'.

SIX

The Whetstone of Witte: content and sources

ULRICH REICH

IN 1557, BARELY A YEAR before his death, Recorde published his most noted work, *The Whetstone of Witte*, in which he presents algebraic ideas in the vernacular. Best known for its introduction of the sign ══════ to indicate equality, the work has a number of other remarkable features, explored in this chapter. The chapter also reveals some of Recorde's sources and compares them with Recorde's own formulations.

Overview

In the Preface to the second book in his earlier work on geometry, *The Pathway to Knowledg*,[1] Recorde had already indicated his intention to publish a book on algebra, which was to include examples 'appertaynyng to the rule of Algeber, applied vnto quantitees partly rationall, and partly surde' (*Pathway*, sig. a.iijᵛ).

The Whetstone of Witte was published in quarto format and contains 164 folios and two folded pull-out pages.[2] The title page indicates the broad content of the work as being 'the seconde parte of Arithmetike: containyng thextraction of Rootes:The *Cossike* practise, with the rule of *Equation*: and the woorkes of *Surde Nombers*'. These terms are explored further within this chapter.

A sixteen-line poem is added, in which Recorde compares his book with a whetstone to sharpen the mind, having prepared his readers in a previous publication: '*The* grounde of artes *did brede this stone*'. He promises that all may benefit from the work, both those of quick intellect and those who are slower learners:

Here if you lift your wittes to whette,
Moche sharpenesse therby shall you gette.
Dull wittes hereby doe greatly mende,
Sharpe wittes are fined to their fulle ende.

The title page ends with an announcement that 'These Bookes are to bee solde, at the Weste doore of Poules, by Jhon Kyngstone', the book's printer. Table 1 provides a breakdown of the content of the book.

Table 1. The number of pages dedicated to each sub-part of *The Whetstone of Witte* (italics indicate titles used in the book)

Content		Pages
Dedication		5
Preface		6
Two poems		1
Section I: Elementary number theory		72
Introduction to diverse kinds of numbers	20	
Of figuralle nombers	52	
Section II: *The extraction of Rootes*		57
Square roots	28	
Of Cubike rootes	24	
Of Compounde rootes	4	
Pull-out page	1	
Section III: *The Arte of Cossike nombers*		131
Of Cossike nombers	86	
The rule of equation, commonly called Algebers Rule	45	
Section IV: *The Arte of Surde nombers*		52

In a five-page Dedicatory Epistle (sigs a.ii–[a.iiij]), typical of its period, Recorde dedicates the book as follows: 'To the right worshipfull, the gouerners, Consulles, and the reste of the companie of venturers into Moscouia, Robert Recorde Physitian, wissheth healthe with continualle increase of commoditie, by their worthie and famous trauell.'

The Muscovy Company was an association of merchant adventurers founded by the explorer Sebastian Cabot in 1555 and given a monopoly of Anglo-Russian trade. The Company also had as one of its aims to search for the North-East Passage. It seems that Recorde advised the Muscovy Company on navigation.

Recorde states that he is sure that his book will be gladly received and goes on to announce his intention to write a book on navigation 'as I dare saie, shall partly satisfie and contente, not onely your expectation, but also the desire of a greate number beside'. The dedication concludes with 'At London the .xii. daie of Nouember.1557.'

The six-page Preface (sigs b.i–b.iii^v) is addressed to 'the gentle Reader'. Recorde describes the importance of numbers and refers to 'the Bookes of Plato, Aristotell, and other aunciente Philosophers' as well as to 'Nicomachus, and diuerse other writers' who declared 'that Arithmetike is the fountaine of all the other [artes]'. He underlines the importance of number to Divinity, Law and Astronomy, 'And as for Physicke, without knowledge and aide of nomber is nothynge.' Finally, Recorde cites Plato on a further five occasions as he describes the connections between number, measure and weight. The Preface is followed by two eight-line poems entitled 'Of the rule of Cose' and 'To the curiouse scanner'. The former plays with the idea of using a variable – the cose – to represent number in the abstract, so that the learner may see 'What thynges by one thyng knowen maie bee'. The latter juxtaposes two types of people, those who 'mende' and those who 'blame', and Recorde exhorts the reader, in prophetic terms, to 'be wise, and learne before, Sith slaunder hurtes it self moste sore'.

The main body of the work is presented in four sections: Part I presents some details of elementary number theory; Part II is a continuation of Recorde's arithmetic *The Ground of Artes*; Part III, the most significant and well-known section, as well as the largest, is a treatise on algebra; and Part IV completes the book with a discussion of surds.

As far as is known, there are no other editions. The author has been able to trace twenty-one original copies of the book in public libraries in the UK,[3] a further eleven copies in the USA[4] and one copy in Germany.[5] Numerous microfilms and other copies of the book also exist.[6]

In this chapter we shall concentrate largely on the algebraic content of the book, including both the introduction of variables (referred to by Recorde as 'Coßike numbers'[7]) and his discussion of equations, which famously includes his introduction of the equals sign. We shall also draw attention to some of the interesting and extraordinary details in other sections of the book.

Recorde's sources

In the sixteenth century, authors were not required to cite the sources of their information. Nevertheless Robert Recorde openly cites on no fewer than forty occasions[8] a range of authors including: Aristotle, Euclid, Nicomachus and

Plato from the classical period; the Roman scholar Boethius (*c.*480–524); and Recorde's own contemporaries, the Italian Cardano (1501–76) and the Germans Stifel (1487–1567) and Scheubel (1494–1570). As we shall see, he also refers twice to the writings of the great Arab mathematician al-Khwārizmi (*c.*780–*c.*850).

The forty citations are distributed unevenly. The Preface refers back to the classical world, quoting Nicomachus once, Aristotle twice and Plato on six occasions. In the main text, apart from one reference to Boethius, Euclid is cited fourteen times and, most significantly, there is one citation from Stifel, five from Cardano and as many as ten from Scheubel.

Cardano[9] was a famous Italian physician and mathematician, best known for his account of the discovery of the rule for the solution of a cubic equation. His most important mathematical work, *Artis magnae, sive de regulis algebraicis* (also known as *Ars magna*), was printed in Nuremberg in 1545. However, Recorde refers, in the five instances, to another of Cardano's works, entitled *Practica arithmetice, & mensurandi singularis*, published in Milan in 1539 (see Figure 1).

Michael Stifel,[10] a priest and early follower of Martin Luther, became well known when he prophesied, on the basis of arithmetical calculation, that the world would end in 1533. When it became evident that his prediction was false, he was arrested and dismissed from his post. He subsequently devoted himself to the study of mathematics. His most important book is his work on algebra, *Arithmetica integra*, published in Nuremberg in 1544 (see Figure 2). The one reference to Stifel occurs in the context of Recorde's introduction of diametral numbers.[11] However it is reasonable to suppose that Recorde also made use of other parts of Stifel's book.

Johann Scheubel[12] was professor of mathematics at the University of Tübingen and published six books on arithmetic, algebra and geometry. He specialized in lecturing on Euclid's *Elements* and was the first to publish parts of the *Elements* (the seventh, eighth and ninth books) in German. Five of the ten references to Scheubel occur when Recorde compares the methods of Cardano and of Scheubel to extract cube roots, and are all taken from Scheubel's arithmetic *De numeris et diversis rationibus*, published in Leipzig in 1545 (see Figure 3). The other five references to Scheubel occur when Recorde discusses cossics and surds, and cites Scheubel's *Algebrae compendiosa facilisque descriptio*, printed twice in Paris in 1551 and 1552 (see Figure 4). It also seems possible that, in his discussion of perfect numbers, Recorde was aware of Scheubel's related work in his German publication of parts of Euclid's *Elements* – *Das sibend, acht und neunt Buch Euclidis* – printed in Augsburg in 1555.

In comparing Stifel and Scheubel, David Eugene Smith succinctly summarizes their attributes and personalities:[13]

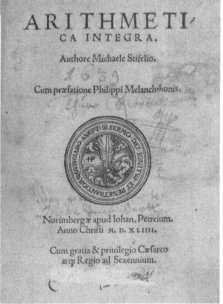

Figure 1. (*above*) Title page
of Cardano's *Practica arithmetice, & mensurandi singularis*
(Milan, 1539). Württembergische
Landesbibliothek Stuttgart, HB 390

Figure 2. (*right*) Title page
of Stifel's *Arithmetica integra*
(Nuremberg, 1544).
Adam-Ries-Bibliothek Annaberg-
Buchholz, ARB-1040-07

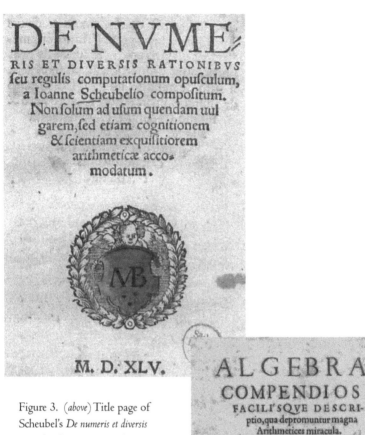

DE NVME꞉
RIS ET DIVERSIS RATIONIBVS
seu regulis computationum opusculum,
a Ioanne Scheubelio compositum.
Non solum ad usum quendam uul
garem, sed etiam cognitionem
& scientiam exquisitiorem
arithmeticæ acco꞉
modatum.

M. D. XLV.

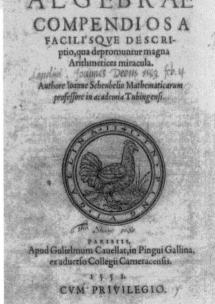

ALGEBRAE
COMPENDIOSA
FACILI'SQVE DESCRI-
ptio, qua depromuntur magna
Arithmetices miracula.

Authore Ioanne Scheubelio Mathematicarum
professore in academia Tubingensi.

PARISIIS,
Apud Gulielmum Cauellat, in Pingui Gallina,
ex aduerso Collegii Cameracensis.

1551.
CVM PRIVILEGIO.

Figure 3. (*above*) Title page of
Scheubel's *De numeris et diversis
rationibus* (Leipzig, 1545).
Sächsische Landesbibliothek – Staats-
und Universitätsbibliothek, Dresden,
Math. 1028

Figure 4. (*right*) Title page of
Scheubel's *Algebrae compendiosa
facilisque descriptio* (Paris, 1551).
National Library of Wales, b51 P3(7)

In marked contrast to Stifel was his contemporary, but slightly his junior, Johann Scheubel. Stifel was brilliant, Scheubel was scholarly; Stifel was popular, Scheubel was heavy; Stifel was eccentric, Scheubel was balanced; and Stifel was effusive, while Scheubel was a man of dignity and poise. The University of Tübingen called Scheubel to a professorship of mathematics at about the same time that Stifel was sent to prison.

Section I: Elementary number theory

Before beginning the main text Recorde repeats the book's headline in a slightly modified form: 'The seconde parte of Arithmetike, containyng the extraction of Rootes in diverse kindes, with the Arte of Cossike nombers, and of Surde nombers also, in sondrie sortes.'

In *The Whetstone of Witte* Recorde continues the style he adopted in *The Ground of Artes*, presenting material via a dialogue between Master and Scholar. This dialogue begins with a comparison between geometry and arithmetic, in which Recorde asserts that, whereas in geometry the basic component is the point, in arithmetic that component is number. He then cites the works of Nicomachus and Boethius as he dedicates twenty pages (sigs A.i–C.iii) to describing 'diuerse kindes' of numbers, introducing a whole raft of terms, many of which are new, including: a *parte* or *partes*, *whole nomber* and *broken nomber* (*commonly called fractions*), *abstracte* and *contracte nombers*, *numerator* and *denominator*, *absolute* and *relatiue nombers*, *figuralle nombers*, *square nombers*, *euen* and *odde nombers*, *compounde* and *simple* (*vncompounde*) *nombers*, *euen nombers euenly* and *euen nombers oddely* (*vneuenly*), *perfecte* and *imperfecte nombers*, *superfluouse* or *abundaunt nombers*, *diminute* or *defectiue nombers*, *odde nombers compounde* and *vncompounde*, *commensurable* and *incommensurable nombers*, *proportion of equalitie* and *inequalitie* (*greater* and *lesser*), *proportion superparticulare* and *superpartiente* and many special cases of these proportions.

An interesting detail is Recorde's treatment of perfect numbers, which he defines as follows: 'Perfecte nombers are suche ones, whose partes joyned together, will make exactly the whole nomber.'[14] Recorde mentions the perfect numbers 6 and 28 and explains the property very exactly as follows:

Likewaies. 28. hath for his halfe. 14. for his quarter. 7. for his seuenth parte. 4. and for his fowertenth parte. 2. and for his. 28. parte. 1. all whiche put together, that is. 1. 2. 4. 7. and. 14. doe make. 28. of this sort there are very fewe more in comparison. And for an example, I sett here, as many as are under. 6000000000. and thei are these. 6. 28. 496. 8128. 130816. 2096128. 33550336. 536854528.

Recorde's list of perfect numbers is not without its errors. The first four – 6, 28, 496 and 8128 – were known to early Greek mathematicians. However, neither of Recorde's next two numbers – 130,816 and 2,096,128 – is perfect. Nor is his last number 536,854,528. He has more luck with the number 33,550,336 which is indeed perfect, as was known to Arabian mathematicians. We cannot be certain of Recorde's source, but it is instructive to compare his statements with those in Scheubel's *Das sibend, acht und neunt Buch Euclidis*. Scheubel also mistakenly includes 2,096,128 and 536,854,528 as perfect numbers. However, in contrast to Recorde, Scheubel realized that 130,816 is not perfect, a conclusion based on one of Euclid's propositions.[15]

The second part of this section of the book (sigs C.iii–[I.iiij]^v) deals extensively (over fifty-two pages) with *figuralle* numbers, by which Recorde means numbers that can be directly related to diagrams or geometrical figures. Recorde defines terms such as *linearie nombers, superficiall* or *flatte nombers* and *bodily* or *sound nombers*, and devotes considerable space to discuss *square nombers, longesquares, diametralle nombers* and *likeflattes*, liberally illustrating his text with helpful diagrams. In his treatment of the ratios of diametral numbers, in which he discusses particular sequences based on the ratio of the two shorter sides in Pythagorean triples, Recorde reveals Michael Stifel (Stifelius) as his source:

> Stifelius doeth set them so, that the numerator standeth for the seconde, or greater side: and the denominator for the firste nomber, or lesser side. And for the most delectable contemplation, to behold their forme or progression, he setteth doune as many whole nombers, as the fraction will giue.

Recorde draws heavily on Stifel's work, replicating precisely the examples of both 'firste order' and 'seconde orde' ratios of diametral numbers listed by Stifel.[16]

At the conclusion of this section Recorde discusses how best to refer to successively higher powers of numbers. In today's terminology, we still refer to the square of a number and the cube of a number, both terms having geometrical resonance. However, for higher powers the language is simplified so that, for example, we may refer to 'the fourth power of 2' or '2 to the power 4', 'the fifth power of 3' or '3 to the power 5'. Recorde seeks to maintain geometrical connotations for the higher powers, and begins (sig. G.iii^v):

> So that one multiplication maketh a *square nomber*
> And twoo multiplications doe make a *Cubike nomber.*
> Likewaies. 3. multiplications, doe give *a square of squares*. And. 4. multiplications doe yelde *a sursolide.*
> And so infinitely.

However, he then introduces alternative terminology, that replaces the term *sursolide* by an alternative geometrical analogy, as illustrated in Table 2, based on a larger version printed by Recorde that includes the integer 'roots' from 1 to 10 rather than only the two roots, 2 and 3, presented here for illustration.

Table 2. A shortened version of a table of powers of numbers set out by Recorde, using his spellings

	The vulgare names	*The table of rooted nombers*		*The authors names*
1	*Rootes*	2	3	*Rootes*
2	*Squares*	4	9	*Squares*
3	*Cubikes*	8	27	*Cubes*
4	*Squares of Squares*	16	81	*Longe Cubes*
5	*Sursolides*	32	243	*Squares of cubes*
6	*Squares of Cubes*	64	729	*Cubike Cubes*
7	*Seconde Sursolides*	128	2187	*Longe Cubike Cubes*
8	*Squares of squared squares*	256	6561	*Squares of Cubike Cubes*
9	*Cubes of Cubes*	512	19683	*Cubes of Cubike Cubes*
10	*Squares of Sursolides*	1024	59049	*Longe Cubes of Cubike Cubes*

Recorde's attempt to popularize the special names *Longe Cubes*, *Longe Cubike Cubes* and *Longe Cubes of Cubike Cubes* proved to be unsuccessful. However, later in this section Recorde introduces even higher powers, amending his suggested terminology and exploiting the medieval Italian word *censo*, which is related to the Latin *census*: '. . . which I doe therfore call *longe Cubes*: but commonly thei bee called *Squared Squares*, or *Squares of Squares*: and of some men thei are named *Zenzizenzikes*, as square numbers are called *Zenzikes*.' Thus, and to prepare us for even more complex expressions, Recorde establishes the term 'zenzike' to mean 'to the power of 2' from which he derives 'zenzizenzike' to mean 'to the power of 4'. Taking a deep breath a few pages later, he continues: 'And so by like reason, doe I cal the nexte nombers *square cubes of cubes*, or *square cubike cubes*: whiche other men doe cal zenzizenzizenzikes, that is *squares of squared squares*.'

This extraordinary word 'zenzizenzizenzike', which therefore means 'to the power of 8', poses a particular challenge in terms of its pronunciation and recurs on two further occasions in the text. Listed in the *Oxford English Dictionary*, with its inclusion in Recorde's work as the only citation, it is commonly believed that this word contains more Zs (a total of six) than any other in the English language.

However, the author has located a further extension of the word by Recorde to indicate 'to the power of 16'. The new word appears on a pull-out page (sig. R.i) as *Zēzizēzizēzizēzikes* (which is an abbreviation of *Zenzizenzizenzizenzikes*) and contains eight Zs!

Section II: The extraction of roots

In this section (sigs K.i–[Q.iiij]ᵛ) Recorde discusses at length how to calculate square roots and cube roots of whole numbers and, very briefly, extends the methods to higher roots. Before exploring his exposition, we may note that, whereas in *The Ground of Artes* and *The Pathway to Knowledg* and in previous sections of *The Whetstone of Witte* he had used phrases such as 'abate. 4. out of. 5.' to indicate the process of subtraction, he now introduces the alternative formulation 'subtracte. 4. out of. 5.', this being the first known use of the word 'subtract' in English.

Recorde begins his treatment of square roots (sigs K.i–N.iiᵛ) by applying a standard method to square numbers, thus ensuring that the answer is a whole number. Thus he demonstrates that the square roots of 5,152,900, 18,766,224, 22,071,204 and 901,740,841 are, respectively, 2,270, 4,332, 4,698 and 30,029. Recorde then chooses the number 296,882 (knowing that the answer is not a whole number) and calculates that the square of 544 is 295,936. He then proceeds to calculate a closer approximation to the answer by adopting the technique of appending six zeros to the original number, noting that: 'if you liste to sette doune. 6. Cyphers before your nomber, you shall not misse $\frac{1}{1000}$ of an vnitie from the true roote'. He therefore sets out to calculate the square root of 296,882,000,000 and thereby finds that a close approximation to his answer is 544,868/1,000.[17] The original thinking shown by Recorde to improve his answer by adding zeros was unusual for his period.

He then proceeds to apply his methods to solve six worked examples in some detail. The first three – 'a question of an armie', 'the seconde question of an armie' and 'the thirde question of an armie' – together with the fifth example – 'a question of encampyng' – involve increasingly difficult calculations on the theme of arranging men in square battalions. The fourth example – 'a question of scalyng' – also has a military setting:

> A citie should bee scaled, beyng double diched. And the inner diche. 32. foote broade. And the walle .21. foote high. The capitain commaundeth ladders to be made of that iuste lengthe, that maie reche from the [ou]ter brow of the inner diche, to the toppe of the wal . . .

A diagram of a right-angled triangle prompts the Scholar, helpfully referring back to Recorde's *The Pathway to Knowledg*, to remark that 'This figure doth occasion me to remember the 33. theoreme of the pathewaie . . .' before setting out Pythagoras' theorem and proceeding to find the answer. In the final example, the Master sets 'a question geographical':

> There be. 2. tounes, as *Chichester* and *Yorke* whiche lye Southe and Northe, and betwene them. 220. miles. A thirde toune as *Excester*, lieth plaine Weste from *Chichester*. 120. miles. I desire to knowe the iuste distaunce of *Yorke* from *Excester*.

The geographical details are remarkably accurate and the Scholar proceeds confidently to apply Pythagoras' theorem again to calculate the answer.

Recorde then goes on to explain how to calculate cube roots (sigs N.iii–Q.iii). As initial examples, he uses the cubic numbers 26,463,592 and 47,832,147 and proceeds to show that their cube roots are, respectively, 298 and 363. Recorde's method of finding the cube root of numbers that are not themselves perfect cubes is particularly interesting. The Scholar asks: 'how shall I dooe to brynge it vnto a fraction, that maie aptly expresse the nigheste roote in that sorte?' to which the Master answers: 'There bee as many waies, as there bee writers almoste.' Recorde cites Cardano's aforementioned *Practica arithmetice, & mensurandi singularis*, in which Cardano sets out his method based on the following approximation: 'In approximatione autem duc radicem in se, deinde duc productum per .3. & quod fit est diuisor superationis.'[18] Recorde translated this text as 'Multiplie the roote squarely, and againe by 3. and that nomber shall be the diuisor vnto the remainer.'

Recorde also knew Scheubel's method,[19] which he translated from the Latin and presented as:

> *Scheubelius* doeth allege an other reason, and inferreth an other order, diuerse from this, and soche as impugneth this, saiyng:
> *Triple the roote, and the square of it also, and adde bothe those nombers together, and.*
> *1. more: And so haue you a denominator for your numeratour.*

Recorde compares both methods by calculating the cube root of 694,582,951 in two different ways. To nine decimal places the exact answer is 885.607677804. Cardano's method predicts the fractional part of the answer to be 1,428,826/2,349,675 (= 0.608095162). By contrast, Scheubel's method predicts the fraction 1,428,826/2,352,331 (= 0.607408566). Recorde calculates his own answer by appending twelve zeros, adapting his method for square roots, and derives the answer to the fractional component as 6,076/10,000

(= 0.6076). Cardano's method is in error by 0.000417, Scheubel's by 0.000269 and Recorde's by 0.000078. Recorde's account does not include a comparison of the accuracy of the results, although he calculated all three.

To make such a comparison, Recorde chooses an easier example based on the challenge in geometrical terms to find the dimensions of a cube having a volume double that of a cube with edges measuring 3 units. The problem, then, is to calculate the cube root of 54, which must be a number between 3 and 4. Here we realize that Cardano's method is equivalent to the application to a particular case of what is now well known as the 'Newton-Raphson method'. Scheubel's approach is equivalent to an application of the method called 'Regula falsi'. Both methods are routinely taught in modern numerical analysis courses.

Cardano's method starts with the value $x_0 = 3$, from which he derives the 'improved' value $x_1 = 3 - \dfrac{27 - 54}{27} = 4$.[20] Recorde comments on this result: '4. whiche is plainly false, for. 4. is the roote of. 64. and not of. 54.' Scheubel's calculation is $x_1 = 3 + \dfrac{54 - 27}{64 - 27} = 3\dfrac{27}{37}$.[21] Recorde writes: 'But by *Scheubelius* rule, it wil be. $3\dfrac{27}{37}$ that is. $3\dfrac{3}{4}$ almoste: whiche is moche nigher the truthe . . . Yet maie it be easily seen, that *Scheubelius* rule is not so good, as he would it were.' Recorde adds six zeros to use his own method and calculates the fractional component to be ' $\dfrac{77}{100}$ and more, by the portion of the remainer, whiche is nighe $\dfrac{1}{700}$ '. He therefore remarks: 'And so dooeth *Scheubelius* rule erre more, then I thought before.'

As the real value of $\sqrt[3]{54}$, to six decimal places, is 3.779763, Cardano's result differs from it by 0.220, Scheubel's result differs by 0.050, and Recorde's result is best with a difference of only 0.0028. We should add that it was well known to Cardano – and also to other mathematicians – that the exactness of a solution can be improved by adding some zeros,[22] so that this technique was not invented by Recorde.

Recorde concludes his discussion of cube roots with five practical examples, including a question concerning the diameter of a gun's mouth, a question based on two cubes and two questions concerning weights. Each example rests on the geometrical principle that, as expressed by Recorde, '*Cubes* . . . doe beare triple proportion, in comparison of their sides,' reminding the reader: 'As you learned before by the. 19. proposition, of the. 8. booke of *Euclide*'.

To complete this section (sigs Q.iii–[Q.iiij]ᵛ) Recorde presents computations involving some higher-order roots. Specifically he finds the 'zenzizenzike roote' (i.e. the fourth root) of 14,641 and of 8,503,056, the 'zenzizenzizenzike roote' (i.e. the eighth root) of 6,561, the 'Cubicubike roote' (i.e. the ninth root) of 512 and of 10,077,696 and finally the 'zenzizenzicubike roote' (i.e. the twelfth root) of 531,441. All the answers are whole numbers.

Section III: The Arte of Cossike nombers

Occupying a total of 131 pages this is the largest and most significant section of the book, in which Recorde explores algebraic concepts.

Cossike nombers

The first sub-section (sigs S.i–[Ee.iiij]) begins with a description of 'Cossike nombers', by which Recorde means variables (or unknowns) and their powers. In modern notation, a single letter (often x) is used to signify a variable, and integer superscripts signify its various powers, enabling us to write x, x^2, x^3, x^4, etc. Such notation had not been developed by Recorde's period. Rather, mathematicians invented elaborate systems of signs to signify different powers.

The first six cossic numbers are indicated by Recorde using the signs ℈, ℞, ℥, ₡, ℥℥ and ℐ℥. In modern terminology we would write, respectively, x^0, x^1 (or, simply, x), x^2, x^3, x^4 and x^5. Recorde's descriptions of these signs are set out in Table 3.

Table 3. Recorde's description of the first six cossic numbers.

Sign	Recorde's description	
℈	Betokeneth nomber absolute: as if it had no signe	x^0
℞	Signifieth the roote of any number	x
℥	Representeth a square number	x^2
₡	Expresseth a Cubike number	x^3
℥℥	Is the signe of a square of squares, or Zenzizenzike	x^4
ℐ℥	Standeth for a Sursolide	x^5

Recorde states that, using only the three signs ℥, ₡ and ℐ℥, he is thereby able to develop a system that describes all higher-order cossic numbers. For example, his sign for x^6 is ℥ ₡ and his sign for x^{12} is ℥℥ ₡. The method fails whenever he meets a power that is a prime number – 5, 7, 11, 13, 17, 19, 23 . . . – which Recorde refers to as sursolide numbers. To overcome this problem he uses the letters A, B, C, D, etc. to label the sursolide numbers. For example, ℐ℥ is the first sursolide (corresponding to x^5, see above), b℥ is the second sursolide (corresponding to x^7) and C℥ is the third sursolide (corresponding to x^{11}). Using this system, any power can be appropriately represented. For example, ℥℥℥b℥ corresponds to x^{56}. Recorde sets out a comprehensive table listing the first eighty-one cossic numbers (see Figure 5).

Figure 5. Recorde's list of the first eighty-one cossic numbers

The author is not aware of such an extensive table being published by any other author. There is an error in the entry for the 69th power, since 69 is not a prime number. The correct sign for the 69th power is ↄ Gſ℥ and, consequently, the entries for the succeeding primes – 71, 73 and 79 – are also in error.

Although Recorde does not indicate his source, it is clear that his system is broadly based on that in Scheubel's algebra (see Figure 6).

SIGNIFICANT AVTEM CHARACTERES,

§ quidem,	Numerum.	2ᵣ, verò	Radicem.
℥,	Quadratum.	ↄ,	Cubum.
℥℥,	Quadratum de quadrato.	ſ℥,	Surſolidum.
℥ↄ,	Quadratum de cubo, vel contrà, Cubum de quadrato.		
Bſ℥,	Biſſurſolidum ſignificat.		
℥℥℥,	Quadratum de quadrati quadrato, vel contrà, Quadratum quadrati de quadrato.		
		cↄ..	Cubum de cubo.
℥ſ℥,	Quadratum de ſurſolido, vel contrà, Surſolidum de quadrato.		
		Tſ℥,	Terſurſolidum.
℥℥ↄ,	Quadratum quadrati de cubo, vel contrà, Cubum quadrati de quadrato.		

Figure 6. Scheubel's system of cossic numbers. *Algebrae compendiosa facilisque descriptio* (Paris, 1551), sig. A.ijᵛ. National Library of Wales, b51 P3(7)

In contrast to Recorde, Scheubel uses the Latin *bissursolidum* and *tersursolidum* and chooses to use the first letters of these words – B and T – to correspond to Recorde's second and third sursolide. Scheubel's list ends with the power 12.

Before Recorde goes on to discuss algebraic operations, he introduces the signs ⊕ for addition and ⊖ for subtraction:

> *Master.* There be other. 2. signes in often vse, of whiche the firste is made thus
> ⊕ and betokeneth more: the other is thus made ⊖ and betokeneth
> lesse.

These signs were invented by Johannes Widmann of Eger (*c*.1460–after 1498) who lectured on algebra at the University of Leipzig. They occur for the first time in manuscripts of the period[23] and are printed in Widmann's arithmetic of 1489.[24] As far as can be ascertained, Recorde was the first to use these signs in any text published in English.

Having introduced these signs, Recorde describes at length the basic operations of addition, subtraction, multiplication and division of algebraic expressions. For example, the method shown in Figure 7 is how Recorde sets out the addition of algebraic terms.

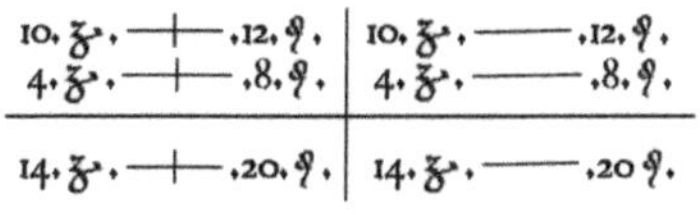

Figure 7. Recorde's method of adding algebraic expressions

We can easily recognize how these expressions can be translated into modern symbols: in the example on the left in Figure 7, Recorde adds the terms $(10x^2 + 12)$ and $(4x^2 + 8)$ to obtain the result $(14x^2 + 20)$ and, in the example on the right, he adds the terms $(10x^2 - 12)$ and $(4x^2 - 8)$ to get $(14x^2 - 20)$.

The multiplication of expressions can also be deciphered in modern terms, although its appearance, as shown in Figure 8, is at first more daunting.

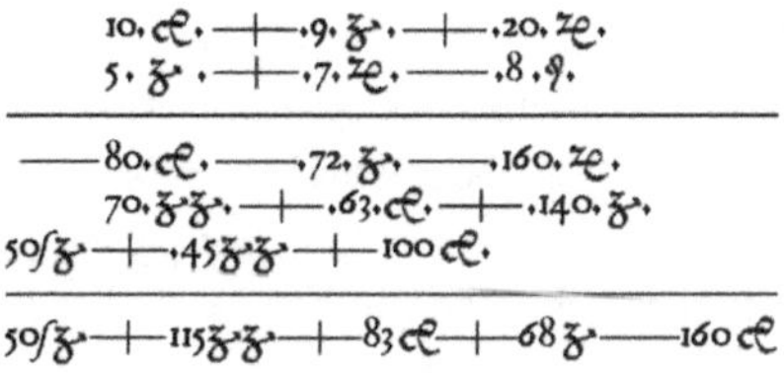

Figure 8. Recorde's method of multiplying algebraic expressions

In this example Recorde multiplies the expressions $(10x^3 + 9x^2 + 20x)$ and $(5x^2 + 7x - 8)$. The correct answer is $(50x^5 + 115x^4 + 83x^3 + 68x^2 - 160x)$, but the printer made a small error by printing $160x^3$ instead of $160x$ in the last line. Recorde then deals with the division of algebraic expressions and devotes a further thirty-five pages to present difficult examples of the basic operations.

At this point in his book an interesting detail reinforces the identity of one of Recorde's sources. During Recorde's period mathematicians were just beginning to grapple with the concept of negative numbers and generally only recognized positive answers to problems as having any real meaning. However, Recorde offers the definition 'An Absurde number expresseth lesse then naught' and presents the example: 'that is. 8. − .12. and is an *Absurde* number. For it betokeneth lesse then naught by. 4.' Michael Stifel was the first to use the word 'absurde' in this context and spoke of zero as '*quod mediat inter numeros ueros et numeros absurdos*'.[25] By contrast, Recorde's normally preferred source, Johann Scheubel, wrote of '*signum priuatiuum vel negatiuum*'.[26]

In the final part of this section on 'Cossike nombers' (sigs [Cc.iiij]–[Ee.iiij]) Recorde presents methods to solve quadratic equations, beginning by classifying such equations into three categories:

> But firste you shall marke, that a Square beeyng compared, as equalle to rootes and nombers, the rootes maie bee coupled with the numbers onely, in. 3. formes. That is. ♆. +. ♀ (whiche is all one with ♀ .+♆.) or els thus. ♀ .−.♆. Or thirdly ♆ − ♀ . And for eche of these. 3. sortes, there is some varietie, in the extraction of the roote. And in them all moche agremente.

We may best explore Recorde's exposition using modern symbols, noting that his classification is equivalent to separating the following three forms:

$$ax^2 = bx + c \ (= c + bx) \qquad ax^2 = c - bx \qquad ax^2 = bx - c$$

Recorde then presents examples of the first form, solving them, with careful attention to providing an explanation, by the method that school pupils would refer to today as 'completing the square'. However, when he considers the equation $x^2 = 4x + 21$ he finds the solution $x = 7$, but misses the negative solution $x = -3$. He proceeds to generalize this particular class of equation to encompass 'Other formes in like sorte', the first such form having the structure $x^{m+2} = bx^{m+1} + cx^m$. Examples of this structure are then provided and solved so that, for example, Recorde shows that the equation $x^3 = 3x^2 + 10x$ has the positive solution $x = 5$. Again he omits any negative solutions and fails to note that $x = 0$ is also a solution. Finally he considers examples having the structure

$x^{2n+m} = bx^{n+m} + cx^m$, illustrating the method with two cases, showing that the equation $x^4 = 80x^2 + 2,000$ has the positive solution $x = 10$ and that the equation $x^6 = 400x^3 + 57,344$ has the positive solution $x = 8$, again omitting negative solutions.

Recorde repeats the process based on the second form $ax^2 = c - bx$, illustrating the method with three examples, again only giving the one positive solution in each case: $x^2 = 60 - 4x$ $(x = 6)$, $x^5 = 162x^3 - 9x^4$ $(x = 9)$ and $x^6 = 275,456 - 26x^3$ $(x = 8)$. His treatment of the third form $ax^2 = bx - c$ is particularly interesting as the five cases he analyses have two positive solutions, both of which he finds. The first two examples each have two positive integer solutions: $x^2 = 16x - 63$ $(x = 7$ and $9)$, $x^6 = 8x^5 - 12x^4$ $(x = 2$ and $6)$. The third example is $x^7 = 2,000x^4 - 470,016x$. Recorde deduces that either $x^3 = 1,728$ or $x^3 = 272$. He finds 12 as the cube root of 1,728 and realizes that the cube root of 272 is not a whole number. The same happens in the next example $x^7 = 12x^4 - 32x$ where he deduces that $x^3 = 8$ or $x^3 = 4$ and that therefore $x = 2$ is an integer solution and the other positive solution is the cube root of 4. The fifth and final example $x^5 = 24x^3 - 135x$ leads to $x^2 = 9$ or 15 and so to the integer solution 3 and also the square root of 15, which is not a whole number.

Recorde concludes this part of his section on 'The Arte of Cossike nombers' with a foretaste of what is to come next: 'And here will I make an eande, of the works of *Cossike* [n]ombers. And now will I applie them to practice in the rule of *equation*, that is commonly called *Algebers* rule.'

The rule of equation

Recorde recognizes that this sub-section (sigs [Ee.iiij]^v–Ll.ii^v) is the highlight of his book:

> . . . now will I teache you that rule, that is the principall in *Cossike* woorkes: and for whiche all the other dooe serue.
>
> This Rule is called the Rule of *Algeber*, after the name of the inuentoure, as some men thinke: or by a name of singular excellencie, as other iudge. But of his vse it is rightly called, the rule of *equation* . . .

The reference here is to the work of the Islamic mathematician and astronomer al-Khwārizmi, the title of whose book in Arabic on solving equations includes the word *al-ǧabr*, which became our *algebra*.

In a section subheaded *The partes of the rule* Recorde begins by commenting on the variety of ways in which equations are classified by other writers and also indicates his own preference:

> This rule of *equation*, is diuided by some men, into diuerse partes. As namely *Scheubelius* dooeth make. 3. rules of it. And in the seconde rule, he putteth. 3. seueralle cannons. Some other men make a greater nomber of distinctions in this rule. But I intende (as I thinke beste for this treatice, which maie serue as farre as their workes doe extende) to distincte it onely into twoo partes. Whereof the firste is, *when one nomber is equalle vnto one other*. And the seconde is *when one nomber is compared as equalle vnto 2. other nombers.*

This is Recorde's first explicit reference to Scheubel's *Algebrae compendiosa facilisque descriptio*. It is probable that the 'other men' are German mathematicians, including Heinrich Schreyber (Grammateus) who defined seven types of equations,[28] and Christoff Rudolff who defined eight types.[29] In some German manuscripts as many as twenty-four distinctions are made.

Before continuing with his exposition Recorde inserts an aside that guarantees his ultimate fame, for it is here, as shown in the original typescript in Figure 9, that we find his invention of the sign of equality as a useful abbreviation: 'And to auoide the tediouse repetition of these woordes: is equalle to: I will sette as I doe often in woorke vse, a pair of paralleles, or Gemowe [twin] lines of one lengthe, thus: = bicause noe. 2. thynges can be moare equalle.'

And to a=
uoide the tediouse repetition of these woordes : is e=
qualle to : I will sette as I doe often in woorke vse, a
paire of paralleles, or Gemowe lines of one lengthe,
thus : ═══ bicause noe. 2. thynges, can be moare
equalle.

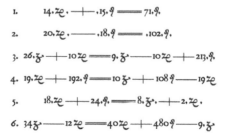

Figure 9. Recorde's explanation of his use of the equals sign

On the same page (see Figure 10) he then presents the first equations ever to be written using what is now a familiar notation.

Figure 10. The first equations to be printed using the equals sign

Today we write these equations as:

1. $14x + 15 = 71$
2. $20x - 18 = 102$
3. $26x^2 + 10x = 9x^2 - 10x + 213$
4. $19x + 192 = 10x^2 + 108 - 19x$
5. $18x + 24 = 8x^2 + 2x$
6. $34x^2 - 12x = 40x + 480 - 9x^2$

Before proceeding to develop his general exposition Recorde pauses to consider these particular equations further in order to illustrate to the reader how the cossic signs and the signs ——|——, ———— and ======= interplay. He explains in detail how to proceed in each case but he only presents a complete solution for the first two linear equations.[30] He rearranges the other (quadratic) equations to forms that are more amenable to solution viz. $17x^2 = 213 - 20x$, $5x^2 = 19x + 42$, $8x^2 = 16x + 24$ and $43x^2 = 52x + 480$.

Recorde then returns to consider the 'varieties of equations' and expands further on his two classifications but not before prefacing his exposition by pointing out that Scheubel's classifications are in fact embedded in his own and by excoriating other needlessly complex systems:

> Now will I shewe you the varieties of equations, taught by *Scheubelius*, bicause you maie perceiue, how thei bee conteined in those .2. formes, named by me. As for the manyfolde varieties, that some other doe teache, I accoumpte it but an idle bablyng, or (to speake moare fauourably of them) an vnnessary distinction.

Recorde starts by presenting his first equation:

> The first equation after *Scheubelius*, & after my meanyng also, is, when one nomber is equall to an other: meanyng that thei bothe must be simple nombers *Cossike*, and vncompounde. As. 6.♂. equalle to. 18.♀

and adding a further six examples. In modern notation these examples are: $6x = 18$, $4x^2 = 12x$, $14x^3 = 70x^2$, $15x^5 = 90x^4$, $20x^6 = 180x^5$, and $26x^{10} = 117x^9$. They are all of the type $ax^{m+1} = bx^m$, where $m = 0, 1, 2, \ldots$ Recorde solves these equations and continues with 'the seconde forme of the firste equation' of the type $ax^{m+n} = bx^m$, where $m = 0, 1, 2, \ldots$ and $n = 2, 3, 4 \ldots$ His three illustrative examples of this second form are: $6x^3 = 24x$, $7x^5 = 567x$, and $7x^5 = 56x^2$.

Recorde continues with his 'seconde kinde of equation', again distinguishing two forms:

> The second kinde of equation, after *Scheubelius* minde and myne also, is, when one simple nomber *Cossike*, is compared as equalle to. 2. other simple nombers *Cossike*, of seueralle denominations, and like distaunce.
>
> And in soche equation, beyng reduced as is taught before, the roote of those. 2. nombers compounded, as in one (or rather the valewe thereof) shal be extracted: As I haue before taughte also. And that roote doeth aunswere to the question.
>
> Howbeeit, here is the like obseruation, as was in the seconde forme of the firste kinde. For if those. 3. denominations be not immediate, but doe omit some other betwene them, then shall you extracte the roote of that laste nomber, in all poinctes, as you did in the firste equation.

To illustrate his reasoning, Recorde presents four examples, all of which have the first form: $4x^2 = 6x + 4$, $6x^5 = 12x^4 + 18x^3$, $5x^5 = 25x^4 - 30x^3$, and $2x^2 = 120 - 8x$. He solves these equations, finding all the positive solutions, including both positive solutions of the third equation. Then Recorde also presents four examples of the second form: $5x^4 = 60x^2 + 320$, $8x^6 = 40x^3 + 30,208$, $8x^6 = 864x^2 - 24x^4$, and $9x^7 = 90x^4 - 144x$. Again he finds all the positive solutions.[31]

We may classify Recorde's equations of the 'seconde kinde', differentiating between those of the first and second forms. There are three cases of the first form, as follows (b and c are positive numbers in all cases):

(R1a) $x^{m+2} = bx^{m+1} + cx^m \; (= cx^m + bx^{m+1})$

(R1b) $x^{m+2} = cx^m - bx^{m+1}$

(R1c) $x^{m+2} = bx^{m+1} - cx^m \quad m = 0, 1, 2 \ldots$

There are also three cases of the second form, which, when n = 1, include the equations of the first form as a special case:

(R2a) $x^{2n+m} = bx^{n+m} + cx^m \; (= cx^m + bx^{n+m})$

(R2b) $x^{2n+m} = cx^m - bx^{n+m}$

(R2c) $x^{2n+m} = bx^{n+m} - cx^m$

Recorde then compares his own classifications with those of Scheubel, his mentor:

But now bicause *Scheubelius* dooeth make. 2. seueralle equations of these. 2. formes: And giueth. 3. diuerse rules, or canons for eche of them, I will declare his. 6. canons to be all contained in this seconde kind of equation.

We may follow Recorde's reasoning by classifying Scheubel's equations as presented in his *Algebrae compendiosa facilisque descriptio*.[32] Scheubel distinguishes three 'canons' within his second equation (a, b and c are positive numbers):

(S1a)	Canon primus	$ax^2 + bx = c$
(S1b)	Canon secundus	$bx + c = ax^2$
(S1c)	Canon tertius	$ax^2 + c = bx$

Then Scheubel introduces a third form of equation, commenting that 'The third equation is nearly the same as the second equation,' and presents examples corresponding to these three types:

(S2a)	$ax^{2n} + bx^n = c$
(S2b)	$bx^n + c = ax^{2n}$
(S2c)	$ax^{2n} + c = bx^n$

We can now see that Scheubel's equations (S1b) and (S2b) are both contained in Recorde's equation (R2a). If we transform (S1a) and (S2a) into the forms $ax^2 = c - bx$ and $ax^{2n} = c - bx^n$, respectively, we see that these equations are both contained in Recorde's equation (R2b). In the same way (S1c) and (S2c) are both contained in (R2c).

In retrospect we witness in this section part of the detail of the laborious process as mathematicians struggled over several centuries to get to grips with equations and their solutions, their efforts often frustrated and encumbered by unwieldy symbolism. Recorde visibly strives to improve the processes, by introducing enabling symbolism, by dismissing the 'idle bablying' of overelaboration, and by diligently searching for simplifications. We sense the excitement as we look over his shoulder.

Recorde follows his discussion of theory by presenting a series of practical applications, in which the Master leads the Scholar, who is challenged with a variety of contexts that require him to apply the various techniques he has been taught.[33] The Master helpfully encourages him, noting that 'There is nothyng better than exercise, in attainyng any kynde of knowlege'. For example, *A double question* poses two challenges:

There are. 2. menne talkyng together of their monies, and nother of theim willyng to expresse plainly his somme, but in this sorte. The nomber of angelles in my purse, saieth the firste manne, mai bee parted into soche 2. nombers, whiche beyng multiplied together, will make. 24. And their *Cubes* beeyng added together, will make. 280. Then, quod the other man. And the like maie I saie of my money, saue that the *Cubes* of the. 2. partes, will make. 539. Now I desire to knowe, what monie eche of them had.

Recorde also includes 'a question of straunge equation' being the equation $x^3 + 8x^2 = 16x + 2,688$. The Scholar of course has difficulties with this cubic equation: 'All that is aboue my cunnyung. For hetherto I have learned noe rule.' The Master provides the solution $x = 12$ and implies that he found it by a process of trial and error. Recorde adds three more higher-order equations, and states one positive solution for each: $8x^6 = 12x^5 + 128$ $(x = 2)$, $8x^6 = 10x^5 + 20x^4 + 400x^3 + 31,250$ $(x = 5)$, and $x^5 = 6x^3 + 8x^2 + 9$ $(x = 3)$.

Recorde rounds off this key section of the book with a promise of more to come, as the Master remarks:

> But of these and many other verie excellente and
> wonderfulle woorkes of equation, at an other tyme I
> will instructe you farther, if I see your diligence ap-
> plied well in this, that I haue taughte you.
> And therefore here will I make an
> eande of *Cossike* nombers,
> for this tyme.

One can but speculate as to what Recorde had in mind and as to what further knowledge he wished to impart. Cardano had published his seminal work on cubic equations in *Ars magna* in 1545 and Recorde may have wished to make this available in English. Recorde's untimely death a year after publication of *The Whetstone of Witte* may well have deprived his readers of further revelations in the field of algebra.

Section IV.   The Arte of Surde nombers

Recorde's final chapter (sigs Ll.iii–[Rr.iiij]^v) is devoted to a discussion of 'surde' numbers. Surds are irrational numbers[34] such as the square root of 2 or the cube root of 5. Recorde's description is similar to that in Scheubel's algebra, including the symbols he chooses to denote square roots and roots of higher oder:

The firste, that is. $\checkmark$ .is customably set, to signifie a *Square roote*. As this. $\checkmark$ .5. betokeneth the *Square roote* of. 5. And $\checkmark$ .12. is the *Square roote* of. 12. Howbeit many tymes it hath with it, for the moare certeintie the *Cossike* signe. $\eth$. And is written thus. $\checkmark \eth$ .20. the *Square roote* of. 20. And $\checkmark \eth$ .56. the *Square roote* of. 56.

The seconde signe is annexed with *Surde Cubes*, to expresse their rootes. As this . $\curlywedge$ .16. whiche signifieth the *Cubike roote* of. 16. And $\curlywedge$ .20. betokeneth the *Cubike roote* of. 20. And so forthe. But many tymes it hath the *Cossike* signe with it also: as. $\curlywedge$. $\mathcal{C}$. 25 the *Cubike roote* of. 25. And. $\curlywedge$. $\mathcal{C}$ .32. the *Cubike roote* of. 32.

The thirde figure doeth represente a *zenzizenzike roote*. As. $\curlyvee$ .12. is the *zenzizenzike roote* of. 12. And $\curlyvee$ .35. is the *zenzizenzike roote* of. 35. And like-waies if it haue with it the *Cossike* signe. $\eth \eth$. As $\curlyvee \eth \eth$ 24 the *zenzizenzike roote* of. 24. and so of other.

The notation appears to be unusual, in that one 'hook' denotes the square root, two hooks denotes the fourth root (the square root of the square root or zenzi-zenzic root) rather than the cube root and, finally, three hooks denotes the cube root. In this context, hooks are a German invention and are first seen in manu-script form over the period 1500–20 and then in print in the books of Rudolff (1525), Stifel (1545 and 1553) and Scheubel (1550, 1551 and 1552).

Recorde devotes as many as fifty-two pages to explaining the application of basic arithmetic operations to surds, essentially following Scheubel's methods as set out in his *Algebrae compendiosa facilisque descriptio*, although the latter's accounts are more concise. Recorde borrows some of Scheubel's Latin terms and introduces some new words to the English language, including *bimedial, binomial, commensurable, residual* and *universal*.

Recorde suddenly interrupts the dialogue about universal numbers and the reader witnesses his wider concerns and a poignant prediction of his own demise:

Master. You saie truthe. But harke what meaneth that hastie knockyng at the doore?

Scholar. It is a messenger.

Master. What is the message? tel me in mine eare Yea sir is that the mater? Then is there noe remedie, but that I must neglect all studies, and teaching, for to with-stande those daungers. My fortune is not so good, to haue quiete tyme to teache.

Scholar. But my fortune and my fellowes, is moche worse, that your vnquietnes, so hindereth our knowledge. I praie God amende it.

Master. I am inforced to make an eande of this mater: But yet will I promise you, that whiche you shall chalenge of me, when you see me at better laiser: That I will teache you the whole arte of *vniuersalle rootes*. And the extraction of rootes in all *Square Surdes*: with the demonstration of theim, and all the former woorkes.

If I mighte have been quietly permitted, to reste but a litle while longer, I had determined not to haue ceased, till I had ended all these thinges at large. But now farewell. And applie your studie diligently in this that you have learned. And if I maie gette any quietnesse reasonable, I will not forget to performe my promise with an augmentation.

> *Scholar.* My harte is so oppressed with pensifenes,
> by this sodaine vnquietnesse, that I can not expresse
> my grief. But I will praie, with all theim that
> loue honeste knowledge, that God of his
> mercie, will sone ende your troubles,
> and graunte you soche reste, as
> your trauell doeth merite.
> And al that loue lear-
> nyng: saie ther
> to. Amen.
> *Master.* Amen,
> and Amen.

Robert Recorde was imprisoned in the King's Bench Prison, Southwark, and died one year later.

Evaluation and postscript

Robert Recorde's *The Whetstone of Witte* was one of the first books on algebra to be published in Europe and the first to be written in the English language. His pioneering work had a major and lasting influence on the teaching of mathematics in Great Britain. Although best known for his invention of the sign of equality, he left a much wider legacy, stemming from his desire to communicate mathematical ideas widely and with understanding. This legacy includes an extensive mathematical vocabulary, largely based on borrowings from Latin and continental European languages, notably Italian, German and French. Some of these words, such as his use of 'zenzi', have since fallen into disuse, while many others have become incorporated into the standard mathematical lexicon.[35] The legacy also extends to Recorde's use of symbolism as part of a process of simplifying mathematical expressions: his use of the sign of equality together with the signs for addition and subtraction for the first time in

English, coupled with use of the cossic signs and signs indicating square and other roots borrowed from Scheubel, established an unprecedented style of mathematical writing. As with his vocabulary, some of this notation stood the test of time, while others (such as the cossic signs) were gradually replaced by more succinct alternatives.

How should we place Recorde's work on algebra in relation to the work of other European mathematicians? His work is chiefly influenced by the German cossists, whose influence extended for over 100 years, from about 1450 to 1560. In algebraic terms, Recorde was one of the leading European cossists.

What further can be said about the books, authored by Cardano, Stifel and Scheubel, that appear to have influenced Recorde's work on algebra and on his likely access to those books? There appear to be at least ten copies of Cardano's *Practica arithmetice, & mensurandi singularis* in UK libraries,[36] and only another eight can be traced by the author in libraries in other countries. One of those copies, now in the library of the University of Tübingen, was owned by Johann Scheubel. The author has traced fourteen copies of Michael Stifel's *Arithmetica integra* in UK libraries[37] and many copies worldwide.

Recorde cites two of Scheubel's works. There are at least four copies of his *De numeris et diversis rationibus* in UK libraries[38] and the author is aware of a further twenty-six copies worldwide, fourteen of which are in Germany. Of more interest, his *Algebrae compendiosa facilisque descriptio* (1551/2) appears to be fairly widespread, a total of sixty-eight copies having been traced, fifteen of which are located in UK libraries.[39] Of the latter, the copy at the National Library of Wales, Aberystwyth has particularly interesting features. It belonged originally to John Dee (1527–1608), who amassed a considerable stock of books and manuscripts at his library in Mortlake.[40] The title page of his Scheubel is signed by John Dee and dated 14 February 1583. In 1680 the volume came into the possession of one Thomas Shaw, and was acquired by the National Library of Wales in 1976.

Dee's handwriting is evident on twenty-four pages of the book, but there are also five notes in another early hand. The most interesting annotation occurs at the point where Scheubel discusses his 'third form of equation' (see Figure 11). The nature of the note shows that its author was well educated in mathematics and had considerable interest in algebra.

The note in Latin may be transcribed as:

> *Semper a medio ad // extremo (aequali) // distantia requiritur. // in istis. et hic pos=*
> *// sis considerationem ha= // bere maximi minimi // et medii, et hac rotam // com-*
> *parare huius ca= // nones ad tres illos // canones in secunda // aequatione descriptos*

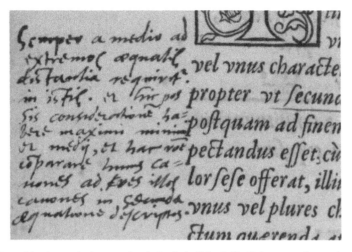

Figure 11. A marginal annotation in Latin in Scheubel's *Algebrae compendiosa facilisque descriptio* (Paris, 1551), sig. F.iij^v. National Library of Wales, b51 P3(7)

and may be translated as:

> You always need the same distance from the middle to the extreme [character]. By these you also can consider the maximum, the minimum and the middle, and here you can compare the rotation of these rules with those three rules described in the second equation.

Is it too fanciful to imagine that this may, indeed, be in Recorde's hand as he began to understand the link between Scheubel's classification of equations and his own, described earlier in this chapter?

Notes

1 See chapter 4.

2 The folios are grouped in fours, the groups labelled consecutively as follows: a, b, A, B, C, D, E, F, G, H, J, K, L, M, N, O, P, Q, R, S, T, U, X, Y, Z, Aa, Bb, Cc, Dd, Ee, Ff, Gg, Hh, Ji, Kk, Ll, Mm, Nn, Oo, Pp, Qq and Rr. The pull-out pages are at sig. R.i and sig. Dd.iii.

3 University of Aberdeen (shelf mark unknown), Cambridge University Library (four copies: LD.42.35, CCD.13.45, SSS.22.5, Peterborough.G.4.14), Cardiff University and Wales NHS Trust Libraries Arts and Social Studies Special Collections (Salisbury WG 30 (1557)), University of Glasgow Main Library (two copies: Sp Coll Ea6-f.12 and

Sp Coll Hunterian R.7.1), British Library, London (two copies: 530.g.37 (MS notes) and 52.a.8), Imperial College, London and Science Museum Library, Swindon (Central Library) (O.B.REC RECORDE), University College London (without shelf mark, acquired 1953), University of London (L.2[B.P.1]SSR), Wellcome Library, London (5364/B), University of Manchester, John Rylands Library (two copies: Deansgate R40576 and Deansgate SC12726A), Bodleian Library, Oxford (two copies: Savile H 12 and Mason H 114), New College Library, Oxford (BT1.135,27), Queen's College Library, Oxford (Sel.f.97) and Tenby Museum and Art Gallery.

4   US Library of Congress (Control No. 6326464, LC Call No. QA33.R32), University of California Berkeley, Cornell University, Ithaca, University of Wisconsin, Madison, Folger Shakespeare Library, Washington, DC (PR1400 20820), Mount Holyoke College, South Hadley, Mass., Yale University, New Haven (QA 101, R 43), John Crerar Library, Chicago, Henry E. Huntington Library, San Marino, Calif., Rare Books (56546) (with reproductions of microfilms), University of Michigan, Ann Arbor (QA 101, R 43) and Columbia University Library, New York (PLIMPTON 511 1557 R24).

5   Bayerische Staatsbibliothek, Munich (BV001634717).

6   For example, a facsimile was produced by the Da Capo Press, Theatrum Orbis Terrarum Ltd, Amsterdam and New York, in 1969.

7   Also written as cossike or cossic numbers, being the variables or unknowns in an algebraic context.

8   Using Recorde's variant spellings, his citations occur as follows in the order in which they appear: sig. b.i: Plato, Aristotell, sig. b.i^v: Nicomachus, Aristotell, sig. b.ii^v: Plato, Plato, Plato, sig. b.iii: Plato, Plato, sig. A.ii: Euclide, Boetius, sig. A.ii^v: Euclide, sig. [A.iiij]: Euclide, Euclide, sig. [F.iiij]^v: Stifelius, sig. G.i^v: Euclide, Euclide, sig. G.ii: Euclide, Euclide, sig. G.iii^v: Euclide, [O.iiij]^v: Cardane, Cardane, Scheubell, Scheubelius, Cardane, sig. P.iii: Cardane, Scheubelius, sig. P.iii^v: Cardanes, Scheubelius, sig. [P.iiij]^v: Scheubelius, Euclide, sig. Q.i: Euclide, sig. Aa.j: Euclide, sig. Aa.j^v: Euclide, sig. Ff.i: Scheubelius, sig. F[f].iii: Scheubelius, sig. F[f].iii^v: Scheubelius, sig. [Ff.iiij]^v: Scheubelius, sig. Gg.i: Scheubelius, sig. [Pp.iiij]: Euclides.

9   Girolamo (or Hieronimo) Cardano or, in Latin, Hieronymus Cardanus. He is also known in English as Jerome Cardan. See also Mario Gliozzi, 'Cardano, Girolamo (1501–76)', *Dictionary of Scientific Biography*, 3 (New York, 1971), pp. 64–7, and Victor J. Katz, *A History of Mathematics: An Introduction* (New York, 1993), pp. 329–37.

10   See Karin Reich, 'Michael Stifel', in Menso Folkerts, Eberhard Knobloch and Karin Reich (eds), *Maß, Zahl und Gewicht*, 2nd edn (Wiesbaden, 2001), pp. 66–89 and David Eugene Smith, *History of Mathematics*, vol. 1 (1923, reprinted by Dover Publications, New York, 1958), pp. 327–8.

11   If the sides of a right-angled triangle are whole numbers (e.g. 3, 4, 5), the product of the two shortest sides is said to be a diametral number (thus $3 \times 4 = 12$ is a diametral number). The term is not in common use.

12 Variously referred to as Johann(es) Scheubel, Scheubelius or Scheybl. See also Ulrich Reich, 'Scheubel(ius), Johann(es)', in *Neue Deutsche Biographie*, vol. 22, ed. Historische Kommission bei der Bayerischen Akademie der Wissenschaften (Berlin, 2005), pp. 709–10.

13 Smith, *History of Mathematics*, p. 329.

14 Using current mathematical terminology, a perfect number is one that is equal to the sum of all its factors, excluding the number itself. For example, the factors of 6 are 1, 2 and 3, as well as 6 itself. Adding 1, 2 and 3 we get 6, the original number. Hence 6 is a perfect number. Perfect numbers continue to play an important role in modern number theory.

15 Euclid had proved that the formula $2^{p-1}(2^p - 1)$ gives an even perfect number when and only when $2^p - 1$ is prime (Euclid, Prop. IX.36). Setting p=9 gives $2^8(2^9 - 1) = 256 \times 511 = 130,816$ which is Recorde's number. However, 511 is not prime, since 511 $= 7 \times 73$.

16 Michael Stifel, *Arithmetica integra* (Nuremberg, 1544), sig. D.iij.

17 Recorde's answer in decimal notation of 544.868 can be compared with the more exact solution 544.86879.

18 Girolamo Cardano, *Practica arithmetice, & mensurandi singularis* (Milan, 1539), sig. D.iiij.

19 Johann Scheubel, *De numeris et diversis rationibus* (Leipzig, 1545), sig. F.iiij^v.

20 The denominator 27 is calculated following Cardano's method by squaring the first approximation 3 and multiplying the answer by 3. The numerator measures by how much the cube of the first approximation (27) differs from the required answer (54).

21 The denominator 37 is calculated following Scheubel's method of trebling the first approximation 3 (to give 9), trebling the square of the approximation 9 (to give 27), adding both results (to give 36) and then adding 1 more, to give the final total of 37.

22 Cardano, *Practica arithmetice*, sig. D.vj.

23 C 80 (1481 and later) and Cod. Leipzig 1470 (from 1486).

24 Johannes Widmann, *Behende vnd hubsche Rechenung auff allen Kauffmanschafft* (Leipzig, 1489).

25 Stifel, *Arithmetica integra*, sig. 249^v.

26 Johann Scheubel, *Algebrae compendiosa facilisque descriptio* (Paris 1551/2), sig. B.iij^v.

27 In Arabic *al-ğabr* means reunion or restoration and refers in the book to the process of balancing the two sides of an equation.

28 Heinrich Schreyber, *Ayn new kunstlich Buech* . . . (1518, printed in Nuremberg in 1521).

29 Christoff Rudolff, *Behend vnnd Hubsch Rechnung* . . . (Strasbourg, 1525).

30 $x = 4$ is the solution to the first equation and $x = 6$ the solution to the second.

31 In the last example Recorde shows that $x^3 = 8$ or $x^3 = 2$, but does not proceed to extract the cube roots.

32 Scheubel, *Algebrae compendiosa facilisque descriptio*, sig. E.iij.

33   The titles include: *A question of ages, A question of debte, A question of pauyng, A question of an armie* (two questions), *An other question of walles, A question of Bricke, A question of a Testament, A question of silkes, A question of money* (two questions), *A question of iorneyng, An other question, A question of proportion, A double question,* and *A question of an armie.*

34   Mathematical terminology distinguishes between rational numbers (the set of all fractions) and irrational numbers, meaning those numbers that cannot be expressed as fractions.

35   The words include *absurd number, abundant* or *superfluous number, algebra* (from *Algeber*), *bimedial, binomial, broken number, commensurable* and *incommensurable, diminute* or *defective number, geometrical progression, perfect* and *imperfect number, root* (in the mathematical meaning), *subtract, surd* and *whole number.* For a discussion of the language of mathematical science in Tudor England see Sonia Piotti, *The First Algebra Printed in English: The Whetstone of Witte (1557) of Robert Recorde* (Milan, 2005).

36   King's College Library, Cambridge (Keynes.Cc02.17), Royal Observatory, Edinburgh (C6.8), Main Library, University of Edinburgh Special Collections (Df.7.68), National Library of Scotland (ICAS.149), British Library, London (1394.a.9.(1.)), Royal Society Library, London, University College Library, London (A-2R8 2S4), Wellcome Library, London (1280/A), Bodleian Library, Oxford (8° C 113 Art.) and St John's College Library, Oxford (G. scam.1.lower.shelf.24).

37   University Library, Cambridge (Rel.c.54.6), Lincoln Cathedral Library, British Library, London (8504.cc.10), Imperial College, London (Science Museum Library), University of London ([DEM] L.1 [Stifel] SSR), University College London, Robinson Library, Newcastle (PI 511 STI), Bodleian Library, Oxford (two copies: Bookstack BB 110 Art. and Savile R13), Christ Church Library, Oxford (OM.1.10), New College Library, Oxford (BT3.180.16), St John's College Library, Oxford (HB4/5.c.3.9), St. Andrews Main Library Special Collections (TypGN.B44PS) and Main Library, University of Edinburgh Special Collections (O.23.22).

38   University Library, Cambridge (two copies: Hhh.748 and CCD.13.64), British Library, London (1393.c.30.) and University Library of Aberdeen.

39   University of Aberdeen (pi 5121 SCH), Cambridge University Library (Syn.7.57.25), Trinity College, Dublin (ee.k.70), University Library, Glasgow (two copies: Sp Coll Ea7-f.6 and Sp Coll Bi7-g.2), British Library, London (two copies: 529.g.8.(3) and 530.g.2), University College London (STRONG ROOM C 1551 S1), University of London ([DEM] L.2 [Scheubelius] SSR), Bodleian Library, Oxford (four copies: A9.9(2)Linc., 4° B 42(8) Art., 4° S 40 Art.Seld. and D 4.19(1) Linc.), Christ Church Library, Oxford (Special Collections f.2.12) and National Library of Wales, Aberystwyth (b51 P3(7)).

40   The copy is described under item number 661 in Julian Roberts and Andrew G. Watson (eds), *John Dee's Library Catalogue* (London, 1990), p. 91.

SEVEN

The Welsh context of Robert Recorde

NIA M. W. POWELL

ROBERT RECORDE, born *c.*1510[1] into a merchant family in Tenby, Pembrokeshire, south Wales, has a well-deserved reputation as a major figure in sixteenth-century learning, whose achievements in the field of science, and in mathematics in particular, were given contemporary recognition.[2] Despite his birth in Tenby, however, perceived influences upon his emergence as a sixteenth-century mathematician have not often included Wales. Quite apart from Recorde's educational experience at the English universities of Oxford and Cambridge, which have been taken to be his primary inspiration,[3] historical interpretation of Wales during the sixteenth century has presented it as an impoverished country, marginal in terms of geography and isolated in terms of cultural interaction, so that the preconditions for a vibrant intellectual sphere, conducive to scientific and mathematical innovation, were absent.[4] The aim of this study is to reconsider and challenge these assumptions. A brief examination will first be made of Recorde's published works in order to identify elements that were of central concern for the author, before evaluating whether, and to what degree, these concerns were forged by his Welsh upbringing and background. In the process, the presumed poverty and cultural isolation of Wales during the early sixteenth century will also be placed under review and reconsidered.

Recorde is probably best known today for having popularized, if not for having devised, the 'equals' sign of two parallel lines in mathematical language, which first appeared in his second volume on arithmetic, *The Whetstone of Witte* published in 1557, but he was also author of a series of influential publications that formed a developing and schematic presentation of mathematics, including

Euclidean geometry. Written in English, and using Hindu–Arabic numerals, they were designed very much as guides to inform and educate with an emphasis on the practical application of mathematical language. Practicality, especially with regard to mercantile practice and navigational arts, was paramount throughout. This was evident in the very first of his mathematical publications, *The Ground of Artes* (1543), a basic introduction to arithmetic and computation, or 'groundwork' as its name suggests, that became what John Denniss and Fenny Smith in chapter 2 call the 'core syllabus of arithmetic texts' for 300 years, running to some forty-five printings.[5] Originally concentrating on subtraction, addition, multiplication and division, and expanded by Recorde himself to include fractions in 1552, elements in this work were evidently aimed at the needs of the contemporary mercantile community operating within a growing fiscal economy. These included weight and money conversion, and methods of apportioning profits. The importance of this pragmatic issue is emphasized by the woodcut on the title page of the 1543 edition, where four men are illustrated in deep discussion over a computational exercise (see Figure 1). Owners of subsequent editions of the *Ground* sometimes highlighted the practical applications with colourful marginal embellishments to delight the reader (for example, see Figure 2).

Practical application was again a paramount consideration in *The Pathway to Knowledg*, first published in 1551, which introduced geometry. Its instructional element was highlighted by the fact that it consisted of a simplification of the first four books of Euclid's *Elements*, now published in the English language for general consumption, but emphasizing the pragmatic utility of geometry by discussing methods of surveying. It is thought that practical, applied geometry was further developed in a text no longer extant, *The Gate of Knowledge*, which may have included discussion of navigational instruments, including the quadrant.[6] The use of mathematical calculation in navigation, as well as in mercantile affairs, was to be a prime concern of Recorde.

Figure 1. Woodcut from the title page of the 1543 edition of *The Ground of Artes*

The Castle of Knowledge, published in 1556, appears at first sight to introduce mathematical issues of greater theoretical concern. It showed how mathematics could be used to articulate astronomy, and that it was a crucial instrument in the Ptolemaic exposition of cosmology. It did, however, reflect Recorde's acquaintance with, and critical assessment of, not only classical and medieval views of cosmology, but also the most recent Copernican theories. More significantly, however, *The Castle of Knowledge* again placed considerable emphasis on practical elements in the second, and particularly in the third, of its four treatises on the material sphere of the world.[7] This complemented the more theoretical elements by giving

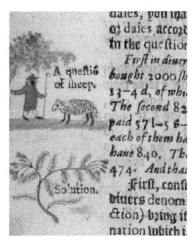

Figure 2. A marginal illustration from a 1623 edition of *The Ground of Artes*. Robinson Library Special Collections, Newcastle University, RB513 REC

practical instruction on how to create a sphere to represent the earth, adapting instructions given in the second treatise on how to fashion a celestial globe, and providing methods for the calculation and measuring of long distances. This, yet again, may well have been developed with the needs of navigation in mind, since it considered the relationship between latitude and the light of day, the shape of the earth and climatic zones. Specific reference is made to instruments of celestial measure commonly used by mariners, including the astrolabe and quadrant, and to the 'names of the thirtye and two pointes in the shippe compasse, whiche bee the Windes names that Mariners sayle by' (*Castle*, p. 51). The origins of Recorde's pronounced concern for navigation, and presenting methods of applying mathematical language and techniques to facilitate naval transport, will be a central concern of this present study.

Practicality and pragmatism in using mathematical language continued to be a feature of *The Whetstone of Witte* (1557), his final publication.[8] This returned further to develop arithmetic, but also introduced for the first time a treatise on algebra that was evidently based on other contemporary work including that of Johann Scheubel of Germany, whom Recorde acknowledged as an authority in the *Whetstone*.[9] A treatise on algebra was certainly not in his mind in 1543 when he published *The Ground of Artes*.[10] It is towards the end of his treatise on algebra in the *Whetstone*, in a section headed 'The Arte of Cossike nombers', that the equals sign is introduced as a symbol, intended as a pragmatic and practical solution 'to auoide the tediouse repetition of these woordes: is equalle to' (*Whetstone*, sig. Ff.iᵛ). He uses the sign for the rest of the section, but reverts to use the more wordy 'is equal to', 'doeth make', or 'that is' – English counterparts for the Latin '*valet*' – in the final section of the book on 'surds', or irrational numbers. This practical device may thus have been a late consideration or an afterthought that was adopted during a stage when the section on algebraic equations may have been revised. It could also be an indication that this may have been the last section to be written. The over-arching explanation given by Recorde of why the simplified symbol for equality should replace the more cumbersome words in his treatment of algebra would be redundant if his initial intention had been to return to use such words in the following section.[11] The introduction of the twin lines as a symbol for equality of value indicates yet again an aspect of practicality in mathematical language. The link between Recorde and the practical application of mathematics is also emphasized by the dedication of *The Whetstone of Witte* to the Muscovy Company founded by John Cabot in 1555, a merchant adventurer company that Recorde appears to have advised on navigational matters, including advice given with regard to the quest for a North-West Passage for navigation during the 1550s. The dedicatory letter has the added interest of offering to produce yet another book specifically to aid practical navigation, illustrating once again Recorde's overwhelming interest in this topic, although such a work was never carried out.[12]

Recorde wrote, then, for merchants, traders and navigators. He lived, of course, in an age of inquiry that boasted both intellectual and technological change, an age claimed to amount to a rebirth or Renaissance. At the most funda-mental level this had led to an inquiry into both language and literature, includ-ing a process of giving the strength of form and definition in both fields. To a large extent, Robert Recorde's mathematical treatises were part of this movement that brought form and definition to language – in his case the language of math-ematics. He also used perhaps the most significant technological achievement of the Renaissance period, the printing press, as a means of communicating this lan-guage to an emerging literate audience. Howell Lloyd and others have argued con-vincingly that he may not have been a pioneer in terms of mathematical theory, but he certainly was in terms of making mathematics understandable and in the creation of a computationally literate public in England and Wales.[13] He himself emphasized this utilitarian aspect and the importance of using the vernacular as a means of communication – and it was English in his case. In his 'declara-tion of the profyte of Arithmetyke', which formed the Preface to the *Ground*, Recorde alluded to the importance of making material understandable when he complained of scorn and derision poured on 'learned men [that] haue taken paynes to do thynges for the ayde of the vnlerned' in England on account of the failure of those learned men to engage with the language of common discourse.

Publishing in the English vernacular was one way he perceived of overcoming this barrier between the learned and unlearned. The success and reissue of his own work in the English vernacular is testimony to the success of that process, and his pragmatic understanding of the importance of communication.[14]

As noted above, despite Recorde's birth in Tenby, south-west Wales, perceived influences upon his emergence as a mathematician have not often included Wales. Anthony Wood opens his treatment of him in his *Athenae Oxonienses* by stating that he 'received his first breath among the Cambrians', but adds that he had no idea in which county he was born, and leaves any Welsh connection at that.[15] In a survey of his life highlighting possible links with other Welshmen, Howell Lloyd concludes that his connections with Wales during his adult life were minimal, although family links with his kindred in Tenby continued to his death and he retained property there.[16] The posthumous editing of his work by John Dee may reflect one specifically Welsh association in London,[17] and yet the origins of his scientific interests tend to be linked more with his university career. An Oxford man, he was admitted as BA on 16 February 1531. His talents were recognized by his appointment as a Fellow of All Souls during the same year, and he was granted a licence to incept as MA at Oxford in 1534. He moved to Cambridge *c.*1535 to study medicine, and was finally granted grace to incept as a Doctor of Medicine at Cambridge early in 1545.[18] During his time at Oxford his reputation as a clear elucidator of mathematics, emphasizing practical application, is noted in particular by Wood in his passage on him, 'He publicly taught arithmetic and the grounds of mathematics, with art of true accompting. All which he rendered so clear and obvious to capacities, that none ever did the like before him in the memory of man.'[19] Wood, however, does not suggest that his mathematical knowledge was acquired other than at Oxford. By 1547, within two years of his graduation at Cambridge, he was evidently in London when his brief text-book on the medical analysis of urine, *The Vrinal of Physick*, was published.[20] This education in the universities of England and then his service to the state in his capacity as Comptroller of the mints in Bristol and Ireland have been considered far more formative in his development as a mathematician than any association with Wales.

Perhaps one should not have expected too much from Wales. In a seminal lecture entitled 'The Renaissance' delivered in 1987 by the doyen of historians of early modern Wales, Glanmor Williams, his central argument was that Wales was a land less than propitious for the absorption of new ideas.[21] Its peripheral geographical position, at the very western edge of Europe and, according to Williams, to the north of the main trade routes of the period, cut Wales off from early modern European influences. As a result it would experience only a 'limited Renaissance'. It was also, according to this interpretation, a peripheral region in terms of economic potential and wealth, lacking the urban foci and higher educational establishments where new cultural and intellectual developments flourished in continental Europe. It is not a coincidence that the intellectual developments associated with the Renaissance of the fourteenth and fifteenth centuries came to flower in areas where there was an accumulation of fiscal wealth. Trade was one factor in the process of capital accumulation and intellectual opportunity, as illustrated by the emergence of science alongside fine art in Flanders and the Low Countries during the early sixteenth century – an area thriving on trade and manufacturing.[22] Trade was important not only in the generation of wealth, but also in facilitating the movement of ideas alongside goods; trading links and trade routes were thus important conduits for ideas. By contrast, Williams emphasized the absence in Wales of urban foci where such fiscal wealth could be found. Although Wales had over a hundred towns, they were, in terms of population numbers, he claimed, mere villages. This meant that Wales lacked an urban centre capable of supporting Renaissance culture. That was not all. Stifled by its limiting geography, the accumulation of wealth from landed interests in Wales was also insufficient, according to Williams, to sustain large courts of substance which were able to provide patronage for cultural and intellectual pursuits. The tendency was for such noble patronage to draw able Welshmen out of Wales to serve in England rather than enriching the cultural background of Wales itself. This portrayal of Wales as a land shackled by its impoverished upland geography, with isolated and conservative communities struggling to survive by subsistence, was reiterated in Glanmor Williams's last great work, *Wales and the Reformation*, published in 1997.[23] The result, according to Williams, in a flourish of geographical determinism, was a people 'tenacious of their ancient ways and suspicious of innovation'.[24] Others concur in this enduring negative image of early modern Wales. Howell Lloyd, for instance, was of the opinion that Wales was dogged by 'unenlightened persistence in familiar ways', that its involvement in the Renaissance was characterized by sporadic efforts by individuals from Wales, but an involvement that may not have had a profound influence within Wales itself.[25]

If noble patronage tended to draw able Welshmen out of Wales to serve in England rather than enriching their native cultural background,[26] there is no evidence of such aristocratic patronage in the case of Recorde. He was, however, certainly drawn from Wales by his education at Oxford and Cambridge.[27] As noted earlier, Recorde himself emphasized his English links, and his English, paternal origins rather than his Welsh roots. His grandfather, Roger, had moved to Tenby from Eastwell, north of Ashford in Kent, sometime in the late fifteenth century.[28] It was service to the English nation that he emphasized in his Preface to *The Ground of Artes* in 1543, underlining his intent 'to helpe my countre men' to raise themselves from 'the infortunate condition of Englond'. He stressed his English identity even further in later works. 'I am an englysh man,' he said, '& therfore may most easely and plainly write in my natyue tonge,' that is, English, in his justification for using that vernacular in the 1547 *The Vrinal of Physick* (Preface, sig. B.ii*), as he used it in his mathematical treatises. His career, moreover, seems to fit in well with the image portrayed of Wales by Glanmor Williams of a country lacking a political centre and the vibrancy of state-led patronage; a country also lacking universities on its own ground, resulting in a paucity of intellectual opportunities for Welshmen within their own land.

Should the achievement of Recorde thus be understood only through English educational opportunity or did his background in Wales contribute in any way to his achievement in the field of mathematics? Contrary to the very negative views of early modern Wales noted above, it will be argued here that intellectual and economic conditions in Wales during the early sixteenth century may not have been as impoverished or as marginalized as has previously been claimed by historians.[29] This brings forward three important considerations suggesting that the Welsh circumstances in which Recorde was brought up may well have influenced if not moulded his later scientific interests.

In the first place, it is clear that the so-called 'traditional' learning of Wales was not ignorant of computation or science. Arithmetical calculation of Easter and the main saints' days within the calendar in the form of a *computus* has a long tradition in western Christianity,[30] and evidence for this in a Welsh context includes several medieval manuscripts in the Welsh tongue that present methods of carrying out the arithmetic task. A fragment of a pre-Norman Welsh *computus* of the early tenth century survives,[31] and methods of calculation are also expressed later in a strict-metre poetic form ascribed to the fifteenth-century poet Dafydd Nanmor (*fl.* 1445–90), called *Y Compod Manuel* or '*Computus* Manual'. The first surviving version of this metrical composition is thought to be in the hand of Dafydd Nanmor himself,[32] and dated to the second half of the fifteenth century, but the fact that it was copied into several other manuscripts during the early sixteenth century illustrates the continuing importance of such a manual for calendar computation during Recorde's youth, and the widespread circulation of such material. A copy was made *c.*1510 by Sir Thomas ap Ieuan ap Deicws, a priest from Llandrillo in Edeyrnion,[33] and in 1527 Elis Gruffydd of Gronant, Llanasa in Flintshire, the 'soldier of Calais' and servant of Sir Robert Wingfield, incorporated some of the material into the chronicle that he was then writing at Wingfield's house in London.[34] Other early to mid-sixteenth-century copies or versions include that of the poet gentleman Ieuan ap Gruffydd ap Llywelyn Fychan of Llannerch near Denbigh[35] and *Llyfr Llelo Gwta*, written in 1552 by an associate of the Welsh polymath, William Salesbury.[36] Manuscripts in the

National Library of Wales's Cwrtmawr collection also contain copies of other works of scientific interest in the Welsh tongue, some of which include astronomic and astrological material, such as *Llyfr Tesni* (Book of Destiny), which contains a table of the cycle of the sun and the moon.[37]

Aristotelianism in scientific thought certainly dominated in Wales during this period, as illustrated by the strangely corrupted version of Aristotle's name in *Y llyvyr addanvones Alesdottlys i alexandyr Mawr o adynabodigayth dynion wrth i Kyrryf*,[38] but there is also evidence of the adoption of newer elements at a relatively early date. Knowledge and espousal of the Hindu–Arabic, or 'algorithmic', numbering system is one example of this process. Sometime between 1422 and 1430 the gentleman poet Ieuan ap Rhydderch of Cardiganshire (1390–*c*.1470) boasted, in a strict-metre ode, of the scientific subjects that he had mastered during his studies at university during the first and second decades of the fifteenth century.[39] Included in the ode are not only references to his readings of Ptolemy and Aristotle and his study of astronomy, but also descriptions of the astrolabe and quadrant.[40] It is in the description of the quadrant that Ieuan ap Rhydderch states that algorithmic figures were used on it, boasting his implied understanding and appreciation of them, 'Eithr ei fod oll, uthraf dim,/O rygraff rif yr awgrim' [Yet, that it consists, most awesomely of all, entirely of precise algorithmic numbers].[41] Ieuan ap Rhydderch may have been a university graduate whose acquaintance with Hindu–Arabic numbers was gained outside Wales, but its acceptance within Wales during the fifteenth century is also evidenced by an intriguing manuscript of 1488/9 in the hand of another Welsh poet and scholar, Gutun Owain, or Gruffydd ap Huw ab Owain (*fl.* 1460–1500). The manuscript contains inter alia a planetary chart and a *computus* for feast days.[42] Its significance lies in the fact that it is rendered in figures rather than in words and, moreover, in Hindu–Arabic rather than Roman numerals (see Figure 3).

Although this numbering system had been introduced into Europe as early as the late tenth century,[43] it was only during the thirteenth century that it began to be promoted more generally and used in mercantile circles,[44] and appears to have gained currency in the British Isles only during the fifteenth century.[45] Chaucer's use of the term 'nowmbres of augrym' in his *A Treatise on the Astrolabe* (*c*.1390) is one of the earliest references in English in the context of Hindu–Arabic numerical forms,[46] and one of the few pre-1400 manuscripts to display the use in practice of the new numbering system was prepared as late as 1386 for John of Gaunt by Nicholas of Lynne.[47] Gutun Owain's use of this numerical system in his 1480s manuscript was thus at the cusp of the introduction of the Hindu–Arabic system into computational discourse within a British context. In the standard context of traditional Welsh learning, Gutun is considered as a major poet from north-east Wales who strove to preserve genealogical, historical and bardic knowledge by

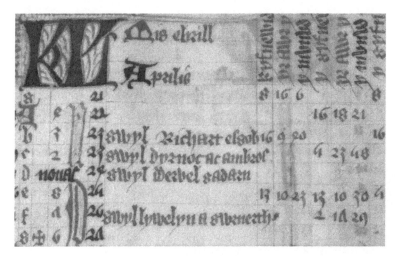

Figure 3. Gutun Owain's use of Hindu–Arabic numerals in 1488/9
in his list of feast days for the month of April (*mis Ebrill*).
National Library of Wales. NLW MS 3026C, p.16

committing it to writing in the generation immediately before Recorde.[48] The
material in this illuminated manuscript contains medical and astrological material,
as well as the *computus*, that has not been generally associated with that traditional
learning, and may have been prepared on commission by an unknown individual
or institution. What is striking here, however, is evidence that such a link existed
between medieval scientific knowledge – including computational practices – and
Welsh learning, and that this went beyond the mere introduction of such ideas at
the initiative of an individual who had experience of university education. Gutun
Owain appears to use Hindu–Arabic numerals with confidence, and their use in
a manuscript written in the Welsh language presupposes a readership that would
understand both.

Recorde, of course, may well not have known this material circulating within
Welsh circles in north-east Wales, but in his *The Ground of Artes* is found an early
explanation of the Hindu–Arabic numeration system. He appears, at the end of his
Preface, to prefer Roman 'figures of nomber' as the symbols that 'muste be learned',
and then uses them in his first section on 'The commodities of Arythmetyke'
(*Ground*, sigs A[.i]–[A.vii]). In the section entitled 'Numeratyon' (*Ground*, sigs
[A.vii]ᵛ–C[.i]) that immediately follows, however, he provides careful instruction
on the use of Hindu–Arabic symbols and the numbering system based on them.
Brought up in Tenby, on the northern shore of the Bristol Channel, it may well be
that he encountered such developments in enumeration through contact with the

city of Bristol itself, or through wider contacts discussed below.[49] Nevertheless, his mother, Rose, was said by the genealogist Lewis Dwnn to have been the daughter of a Thomas ap Sion of Machynlleth, whose form of address could indicate the Welshness of her background. Further links with Welsh families and their cultural world were cemented by marriage between members of the Recorde family and families of the middle march of Wales later in the sixteenth century.[50] Whether Robert Recorde himself had direct links with such circles or not, the Gutun Owain manuscript does, indeed, indicate the general intellectual ambience of the period in Wales. It indicates awareness not only of Aristotelian science but also of newer methods. Such knowledge was in circulation, and challenges the notion of a one-dimensional literary culture for Wales that was divorced from external influences during the age into which Robert Recorde was born.

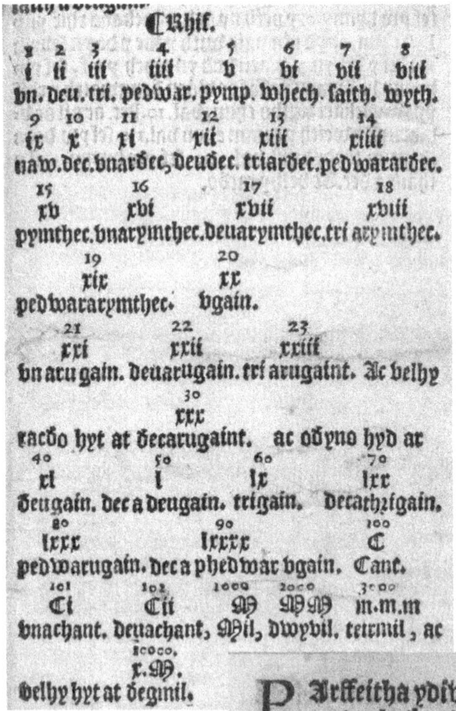
Rhif.
1 2 3 4 5 6 7 8
i ii iii iiii v vi vii viii
vn. deu. tri. pedwar. pymp. whech. saith. wyth.
9 10 11 12 13 14
ix x xi xii xiii xiiii
naw. dec. vnarddec, deudec. triarddec. pedwararddec.
15 16 17 18
xv xvi xvii xviii
pymthec. vnarymthec. deuarymthec. tri arymthec.
19 20
xix xx
pedwararymthec. vgain.
21 22 23
xxi xxii xxiii
vn arugain. deuarugain. tri arugaint. Ac velly
30
xxx
racdo hyt at decarugaint. ac oddyno hyd ar
40 50 60 70
xl l lx lxx
deugain. dec a deugain. trigain. decathrigain.
80 90 100
lxxx lxxxx C
pedwarugain. dec a phedwar vgain. Cant.
101 102 1000 2000 3000
Ci Cii M MM m.m.m
vnachant. deuachant, Mil, dwyvil. teirmil, ac
10000.
x.M.
velly hyt at degmil.

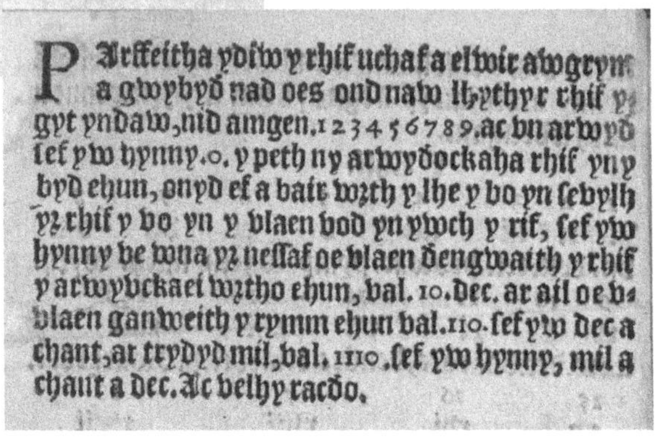
PArffeitha ydiw y rhif uchaf a elwir awgrym a gwybydd nad oes ond naw llythyr rhif yn gyt yndaw, nid amgen. 1 2 3 4 5 6 7 8 9. ac vn arwydd sef yw hynny. o. y peth ny arwyddockaha rhif yny vyd ehun, onyd ef a bair wrth y lhe y bo yn sevyll yr rhif y bo yn y blaen bod yn ywch y rif, sef yw hynny ve wna yr nessaf oe blaen dengwaith y rhif y arwyddockaei wrtho ehun, val. 10. dec. ar ail oe v blaen ganweith y rymm ehun val. 110. sef yw dec a chant, at trydyd mil, val. 1110. sef yw hynny, mil a chant a dec. Ac velly racdo.

Figure 4. Sir John Prise's list of numbers as set out in *Yny Lhyvyr Hwnn* (1546). On the left he uses both Hindu–Arabic and Roman notation and adds the Welsh words for the numerals, based on the traditional vigesimal system. Below he explains the basis of the Hindu–Arabic system.
National Library of Wales, NLW MS WS55, pp. 23–4

Engagement with the Hindu–Arabic numbering system within a Welsh context among contemporaries of Recorde is also revealed in a startingly clear exposition of it in the very first book printed in the Welsh language, Sir John Prise's *Yny Lhyvyr Hwnn* (1546).[51] Although *Yny Lhyvyr Hwnn* has been interpreted in the past as a 'handbook of elementary religious instruction',[52] the main body of the book sought to provide an array of useful secular information for contemporary Welshmen in their own tongue, including advice on orthography and a monthly calendar. Following this, and before printing a text of the creed, the *pater noster* and the ten commandments, are two pages dedicated to an explanation of the Hindu–Arabic system of numeration, including a table of figures showing corresponding Hindu–Arabic, Roman and the equivalent figures in Welsh words. It goes well beyond the figures one to nine discussed by Recorde in *The Ground of Artes*, giving all figures up to twenty-three, then all multiples of ten to one hundred, one hundred and one and two, then a thousand, two and three thousand and lastly ten thousand. Although Prise uses only Roman and word numeration methods in the body of his book, he recommends the Hindu–Arabic system as the most perfect, 'Perffeithia ydiw y rhif uchaf a elwir awgrym' (The most perfect is the topmost number, which is called algorithm) (see Figure 4). The explanation of place value that follows Prise's table of figures is stunningly clear. Prise, like Recorde, was university-educated, but his use of the word 'awgrym', already used in poetry in the early fourteenth century, indicates that the system was already known in Welsh circles; and its promotion by Prise as useful, if not essential, information for contemporary Welshmen reflects its recognition already in learned Welsh circles by the second quarter of the sixteenth century as an optimal method of expression.[53]

The second consideration with regard to Recorde's Welsh background is that practical computation was a very necessary skill in the early sixteenth century in an increasing variety of tasks, both private and public, in Wales as elsewhere. The assessment and collection of dues, taxes and mises during the later medieval period certainly illustrates this well, as it required such expertise, and central government records show the familiarity of local officials in Wales with computational techniques in carrying out such tasks.[54] Records indicate, however, that during the first half of the sixteenth century, Roman enumeration continued to be used without adopting the less cumbersome Hindu–Arabic numbering that had already been used by Gutun Owain half a century earlier. The following example[55] shows how fiscal records were drawn up during the 1540s, in response to the first lay subsidy that was ever granted in Parliament for Wales as a whole in 1543,[56] the very year in which Recorde published his *The Ground of Artes*. All the 1543 records returned to the Exchequer use the cumbersome Roman numerals, and the document relating to the first collection of the tax at Wrexham

Figure 5. Use of *valet* to signify 'equal to' in the assessment for
the first collection of the 1543 subsidy for Wrexham.
The National Archives, E 179/220/166

(see Figure 5) illustrates also the use of '*valet*', taken from the Latin to mean 'equal to' or 'having the value'.

Conservatism in contemporary accountancy may explain why Recorde's own examples of mercantile accounting in the *Ground* sometimes use Roman numerals, despite using Hindu–Arabic forms elsewhere.[57] He himself alludes to the general reluctance to adopt new methods, 'I see moare menne to acknowledge the benifite of nomber, then I can espie willyng to studie, to attaine the benifites of it. Many praise it, but fewe dooe greatly practise it' (*Whetstone*, sig. b.iii);[58] but the change to Hindu–Arabic numerals and simplification of mathematical symbols, as recommended by Recorde, soon appears in other numerical data. Surviving contemporary ecclesiastical documents, relating to information requested by the Privy Council in July 1563 from each diocese in the provinces of Canterbury and York about the number of households in each parish, the so-called Bishops' Census, raise a tantalizing question as to the influence of the area in which Recorde was brought up on this spread of Hindu–Arabic numerals in collating such data. The returns are overwhelmingly in Roman numerals for the province of Canterbury,[59] with the diocese of St David's, in which Tenby lies, standing out as an exceptional suffragan diocese where all the returns are given in Hindu–Arabic numerals.

By contrast, surviving returns for the province of York are in Arabic numerals. The archbishop of York at the time the information was collated in 1563 was Thomas Young, born some three years before Robert Recorde and brought up within five miles of Recorde at Hodgeston, Pembrokeshire. He had been precentor of St David's from 1542 until he became an exile during the Marian period, returning to become bishop of St David's in 1559[60] and then translated to York. It is within reason to suggest that Young had a formative influence on the use of Arabic numerals both at St David's and then at York, where he became archbishop in 1561. Like Recorde, he had received an Oxford education, where he may well have gained knowledge of the new system of enumeration, but its adoption and pragmatic use by both Young and Recorde may equally suggest that they had both first encountered it in their home area in Wales before they entered university, although there is no way of confirming this. Perhaps it should be added that John Prise, author of *Yny Lhyvyr Hwnn*, was, like Young and Recorde, brought up in the diocese of St David's, but in the town of Brecon, which is not in close proximity to Tenby. Whatever the formative influence, later returns of parish data, particularly in 1603, illustrate the further spread of the pragmatic-ally considered decision to adopt Arabic numerals, aided no doubt by Recorde's essays and manuals on arithmetic computation. For instance, returns of the dioceses of Bangor and Lincoln, which had been in Roman numerals in 1563, were in Hindu–Arabic numerals by 1603.[61]

An additional consideration is the perception of poverty in Wales. Taxation evidence, in itself, poses a significant challenge to the traditional interpretation of an impoverished and backward Wales. Taxation records for Tenby, for instance, the birthplace of Recorde, indicate a vibrant and wealthy merchant community in the early sixteenth century, where fiscal computation and records would have been an everyday necessity, a community that would have been well served by Recorde's publications.[62] As many as 7.4 per cent of those taxed in the first collection of the parliamentary subsidy granted in 1543[63] were assessed in the highest possible bracket, having goods of over £20 in value, and a further 4.3 per cent were assessed with goods over £10. Recorde's own brother, Richard, was in that bracket. Assessed at £18, a considerable fortune in terms of tax assessment, and owning personal property valued at, perhaps, around half a million pounds in early twenty-first-century terms, he lived in the West ward of the town, where no fewer than 16.9 per cent of those taxed were assessed at over £10. Here, therefore, was found a concentration of people of considerable fiscal wealth, who were operating very much within a mercantile urban centre in Wales, and a port at that.

It was, moreover, a community of shipowners, merchants and professionals, including attorneys versed in the laws of the sea. In 1537, William Hutton of Tenby, for instance, acted as attorney in a petition to Thomas Cromwell relating to restitution.[64] Fifteenth-century evidence of ships and cargoes owned by the town's merchants confirms its wealth. In 1405, three ships carrying cargo valued at £970 for five Tenby merchants, Thomas Nony, John Banowe, William Score,

John Gwen (Owen?) and John Vaghan, were seized in Spain. Later in the century, in 1481, the *Trinitie*, a Tenby ship, was driven ashore in Cornwall, laden with 130 tuns of osay wine, with five tuns of wax, sugar and oil valued at £1,200.[66] This indicates the tremendous capital outlay by Tenby merchants throughout the fifteenth century.

This wealth acquired through trade was also reflected in the status and privileges of the town. Grants of quayage – the right to raise money from ships that landed there so as to maintain the quay and town walls – were constantly renewed,[67] while, in 1423, the town's ships were granted exemption from similar charges at Bristol.[68] The town's charter, which granted to the burgesses a high degree of freedom from seigneurial control, was also constantly reviewed and expanded to improve the trading infrastructure.[69] This was the prosperous mercantile community that attracted Robert Recorde's grandfather, Roger, to Tenby from Kent.[70] Tenby was thus attractively prosperous for a migrating merchant class. In dealing with their affairs such merchants put a heavy emphasis on literacy and, more importantly, numeracy; indeed, later in the sixteenth century, merchants such as Valentine Broughton in Wrexham, Richard Clough in Denbigh and John Beddoes in Presteigne were prominent in the process of endowing grammar schools.[71] Within mercantile communities such knowledge is thought to have been imparted to youths, prior to the founding of endowed educational establishments, on an informal tutoring basis, although definite evidence of this is not available, and may well have been part of Recorde's own upbringing in Tenby shared, possibly, by Thomas Young.

There is, however, a third consideration. As noted earlier, trade throughout Europe led to external relations and links with distant places. The existence of strikingly wide-ranging trading links was certainly true in the case of Tenby, and this again stands in stark contrast to the negative perceptions and interpretations of an introspective Welsh experience. Reference has already been made to its ships returning from Spain, and the 1481 ship laden with Mediterranean goods.[72] Already in 1403 the 'burgesses, merchants and mariners of the town of Tynbygh' were given protection to trade with Aquitaine, England and Ireland,[73] but early sixteenth-century evidence indicates landings by Breton[74] and, more interestingly, Portuguese sailors.[75] One Portuguese captain, Peter Alves, hired a Tenby mariner, William Phillip, to pilot his ship.[76] What is even more significant is that, in 1537, Thomas Morice, a merchant of nearby Pembroke, was reported by none other than Rowland Lee, then Lord President of the Council in Wales and the Marches, to be 'expert in the Portuguese language',[77] indicating that trading interchange was intensive enough to lead to language transmission in the case of at least one Welshman. This, of course, was the period of burgeoning Portuguese naval exploration, with Vasco da Gama's sailings to the Far East, and then Portuguese travel to Brazil.[78] Peter Alves was said to have been one of a number of Portuguese masters who came to Tenby. Described as a mere 'mariner', his ship was said to belong to the newly created Duke Ferdinand of Aveiro, grandson of John II, the navigator king of Portugal.[79] Ferdinand was involved in naval enterprises in Portugal, and linked with Setúbal, a port in southern Portugal associated with da Gama himself. The presence of such Portuguese mariners in Tenby and Pembroke during the early sixteenth century is significant. Whether any of them were versed in the advanced navigational techniques developed in Portugal at this time is not known for certain, of course, but the transmission of some such knowledge is a strong possibility in view of their association with Ferdinand, and with Setúbal, and the close and direct links between Pembrokeshire and Portugal. Recorde certainly refers to the maritime exploits of Portuguese mariners in his work, and to methods adopted by them to measure distances.[80] It poses the further intriguing question of whether Recorde's own interest in applied mathematics particularly for navigational purposes, which he emphasized so heavily in his treatises, was aroused by these foreign mariners that came ashore in Tenby during his youth. There are indications that Recorde, as a boy, did have particularly strong links with the port of Tenby, since one of the properties recorded in 1586 as part of the Recorde inheritance in the town was a 'Tenement bilt conteyning a garden' located in Crackwell Street, 'at the ending of the wale above the haven against Cornish Clyffe'. This was the street on the cliff edge at Tenby overlooking the port itself, and with easy access to it.[81] The daily circumstances of living in a community that consisted not only of merchants, but also of international mariners, must have been an impressionable experience and must, indeed, have been an all-important factor in Recorde's concern, if not obsession, expressed even in his last published mathematical treatise, for serving the English, if not Welsh, maritime community.[82]

Recorde's home was thus open to European influences – Tenby was on a trade route, it was prosperous and its people had confidence, and such conditions in Wales were not unique to Tenby. Spencer Dimmock has shown a similar pattern of overseas trade and influences in Haverfordwest and Chepstow;[83] and Newport's links with Bristol display a similar pattern;[84] likewise for the north, A. D. Carr has outlined comparable external links for late medieval Beaumaris.[85] Wales, then, was open to external influence, and trade reaching its urban foci can be considered to be both an important stimulant and a conduit for cultural exchange that did not necessarily arrive second-hand from England, as is so frequently presumed. Thus, despite Recorde's insistence on his 'Englishness', it appears that his interest in mathematics, especially applied mathematics, and his specific concern with the use of mathematical calculation in navigation, could well be the product of his own background in Wales, and in Tenby in particular. His pragmatic concerns addressed the needs of the very communities of merchants and navigators that he encountered in early sixteenth-century Tenby. Recorde's Welsh background, then, contributed in no small degree to the structure of his mathematical treatises, and to his later achievements.

Notes

1   Joy B. Easton, 'On the date of Robert Recorde's birth', *Isis*, 57/1 (1966), 121. For his ancestry see Lewys Dwnn, *Heraldic Visitations of Wales and Part of the Marches*, vol. 1, ed. Samuel Rush Meyrick (Llandovery, 1846), pp. 68–9.

2   For an analysis of his reputation, both contemporary and in retrospect, see Howell A. Lloyd, '"Famous in the field of number and measure": Robert Recorde, Renaissance mathematician', *Welsh History Review*, 20/2 (2000), 254–82.

3   See p. 127.

4   See pp. 127–8.

5   The first posthumous, enlarged edition was carried out in 1561 by another Welshman, John Dee: *The Grounde of Artes... Made by M. Robert Recorde... and now of late ouerseen & augmented with new necessarie Additions. ID* (London, 1561).

6   See chapter 5 by Stephen Johnston in this volume. See also Lloyd, 'Famous in the field', p. 264, n. 35.

7   'Wherin is briefly tavght the vse of the Sphere, for certaine conclusions of daily appearaunces and other lyke matters' (*Castle*, p. 61ff.).

8   The *Whetstone* may have been an afterthought to the system that reached its schematic peak in *The Castle of Knowledge.*

9   Scheubel is quoted liberally in the *Whetstone*, particularly in the section entitled 'The rule of equation, commonly called Algebers Rule'. As discussed in chapter 6, Recorde may also have been influenced by the work of Michael Stifel.

10   *Ground*, final page of Preface (unnumbered).

11   An alternative explanation is that he intended to use the symbol only in his discussion of algebra, and to continue to use 'word' devices in his arithmetical discussion of surds. This appears to be a less convincing explanation.

12   *Whetstone*, Dedicatory letter to the Muscovy Company, 'I will . . . shortly set forthe soche a booke of Nauigation' (sig. a.iii).

13   Lloyd, 'Famous in the field', pp. 264–75.

14   See the Bibliography for a full list of the editions of Recorde's works.

15   Anthony à Wood, *Athenae Oxonienses: An Exact History of All the Writers and Bishops who have had their Education in the University of Oxford*, vol. 1 (new edn with additions by Philip Bliss, London, 1813), cols 255–6.

16   Lloyd, 'Famous in the field', pp. 257–9 for his links with John Dee and William Thomas.

17   See n. 5 above.

18   A. B. Emden, *A Biographical Register of the University of Oxford AD 1501–1540* (Oxford, 1974), p. 440; John Venn and J. A. Venn, *Alumni Cantabrigienses*, Part 1: *From the Earliest Times to 1751*, vol. 3 (Cambridge, 1924), p. 435.

19   Wood, *Athenae Oxonienses*, col. 255.

20   *Vrinal.* Recorde's dedication to the Wardens and Company of Surgeons of London, 8 November 1547, was written 'At my house in London'.

21   Glanmor Williams, 'The Renaissance', in *idem* and Robert Owen Jones (eds), *The Celts and the Renaissance: Tradition and Innovation. Proceedings of the Eighth International Congress of Celtic Studies 1987* (Cardiff, 1990), pp. 1–16.

22   K. van Berkel, A. van Helden and L. Palm (eds), *A History of Science in the Netherlands: Survey, Themes and Reference* (Leiden, 1999); H. J. Cook, *Matters of Exchange: Commerce, Medicine and Science in the Dutch Golden Age* (New Haven, 2007). See also Jan Bloemendal and Chris Heesakkers (eds), *Bio-bibliografie van Nederlandse Humanisten* (*Biography and Bibliography of Dutch Humanists*), digital publication, DWC/Huygens Instituut KNAW (The Hague, 2009), *www.dwc.knaw.nl* (accessed 14.09.11); Hendrik Leustra and Alex van den Brandhof (eds), *Biografisch Woordenboek van Nederlandse Wiskundigen* (*Biographical Dictionary of Dutch Mathematicians*), digital publication, DWC/Huygens Instituut KNAW (The Hague, 2009), *www.dwc.knaw.nl* (accessed 14.09.11).

23   Glanmor Williams, *Wales and the Reformation* (Cardiff, 1997). See also *idem*, *The Welsh and their Religion: Historical Essays* (Cardiff, 1991), esp. ch. 5, 'Religion and Welsh literature in the age of the Reformation', pp. 138–72.

24   Williams, *Wales and the Reformation*, pp. 32–3.

25   Howell A. Lloyd, *The Gentry of South West Wales, 1540–1640* (Cardiff, 1968), pp. 204–5, 211–12.

26   Patronage given, for example, to the Thelwall family of Bathafarn Park near Ruthin by the Countess of Warwick during the late sixteenth century. See R. T. Jenkins et al. (eds), *The Dictionary of Welsh Biography down to 1940* (London, 1959), pp. 932–3.

27   See p. 127.

28   Dwnn, *Heraldic Visitations*, p. 68.

29   N. M. W. Powell, '"Near the margin of existence"? Upland prosperity in Wales during the early modern period', *Studia Celtica*, 41/1 (2007), 137–62; *eadem*, 'Do numbers count? Towns in early modern Wales', *Urban History*, 32/1 (2005), 46–67.

30   Alden A. Mosshammer, *The Easter Computus and the Origins of the Christian Era* (Oxford, 2008).

31   Cambridge University Library Additional MS 4543; Ifor Williams, 'The Computus fragment', *Bulletin of the Board of Celtic Studies*, 3/4 (1927), 245–72.

32   Aberystwyth, National Library of Wales, Peniarth MS 52, ff. 40–4 *Y Compod Manuel o waith D[afyd]d Nanmor allan o hen lyfr memrwn.* See also NLW , Cwrtmawr MSS 244B, 206B and 298B for later copies.

33   NLW , Peniarth MS 127, ff. 221–7.

34   Cardiff Central Library, MS 3.4, *The Book of Elis Grufydd* (1527), f. 6.

35   Bodleian Library, Oxford, Welsh MS f. 2, ff. 66r–71v. See also NLW , Peniarth MS 53, pp. 74–87 and Cardiff Central Library, MS 2.4, p. 171 for later copies of the same material.

36   British Library, Add. MS 14,986 (Caer Rhun 9). See also N. M. W. Powell, 'Robert ap Huw: a wanton minstrel of Anglesey', *Welsh Music History*, 3 (1999), 5–29, esp. pp. 6–7 on Salesbury.

37   NLW , Cwrtmawr MS 6B, a 1691/2 copy of earlier material. See also NLW , Cwrtmawr MSS 210B, 672A, 107B and 327B; and NLW , MS 21943, ff. 31–6 for eighteenth- and nineteenth-century copies of earlier material on astronomy and astrology.

38   Cardiff Central Library, MS 3.4, *The Book of Elis Grufydd* (1527), ff. 155b–61 (The book that Aristotle sent to Alexander the Great to come to know men by their bodies). This is a highly abridged version in Welsh of the Greek treatise on physiognomy, a work of uncertain authorship but attributed to Aristotle, which links physical attributes to character and behaviour. The text of the classical *Physiognomonica* has no direct reference to Alexander the Great, although Aristotle was known as his tutor. See W. S. Hett, *Aristotle: Minor Works*, Loeb Classical Library, 307 (London and Cambridge, Mass., 1955 edn), *Physiognomonica*, I–VI, pp. 84–137. There are several later manuscript copies of the Welsh abridgement, including those in British Library, Add. MS 14,979 (sixteenth century) and NLW , Cwrtmawr MS 327B, pp. 32–40, entitled *Llyfr Alesdotlys.* The strangest corruption of all of Aristotle's name appears in a further, eighteenth-century abridgement entitled *Llyfr o waith Ellis Totlis i Alexander Fawr i adnabod Corph Pob Dyn* (Aristotle's book for Alexander the Great to understand the human body), in NLW , Cwrtmawr MS 492B, pp. 41–8.

39   Ieuan ap Rhydderch, 'Cywydd y fost' (An Ode of Boast), in R. Iestyn Daniel (ed.), *Gwaith Ieuan ap Rhydderch* (Aberystwyth, 2003), pp. 50–64. On Ieuan ap Rhydderch's career and dates, see ibid, pp. 4–9, 15, 17, 29–33, 139–44. *Awgrym/awgrim* was used during the first half of the fourteenth century by the poet Dafydd ap Gwilym, but he seems to have been referring to calculation by counters rather than Hindu–Arabic numerals as such ('Morfudd yn edliw ei lyfrdra', in Ifor Williams and Thomas Roberts, *Cywyddau Dafydd ap Gwilym a'i Gyfoeswyr* (Cardiff, 1935), pp. 30, 182–3. Later use of the word by the poets Lewis Glyn Cothi (*c*.1420–90) and Tudur Aled (*fl*. 1480–1525) adopts the derivative meaning of symbol (see T. Gwynn Jones (ed.), *Gwaith Tudur Aled* (Cardiff, 1926), pp. 507, 645). Ieuan ap Rhydderch is the first to use the word to indicate the numerical symbols.

40   Daniel, *Gwaith Ieuan ap Rhydderch*, 'Cywydd y fost', pp. 50–2, lines 29–74.

41   Ibid., p. 51, lines 59–60 (author's translation).

42 NLW, MS 3026C (Mostyn 88), pp. 10, 13–24. The manuscript also contains a treatise on urine, pp. 28–36. See also Morfydd E. Owen, 'Prolegomena i astudiaeth lawn o lsgr. NLW 3026, Mostyn 88 a'i harwyddocâd', in R. Iestyn Daniel et al. (eds), *Cyfoeth y Testun: Ysgrifau ar Lenyddiaeth Gymraeg yr Oesoedd Canol* (Cardiff, 2003), pp. 349–84.

43 Said to have been introduced into Moorish Spain through trading links with North Africa, the earliest appearance of Hindu–Arabic numerals in Europe is in the Codex Vivilanus of 976, El Escorial Library, Escorialensis MS d I 2. Later medieval copies of Robert of Chester's 1145 Latin translation of Musa al-Khowārizmi's *Algebra*, carried out when he was in Segovia, contains Roman numerals. See L. C. Karpinski (ed.), *Robert of Chester's Latin Translation of the Algebra of al-Khowārizmi* (New York, 1915), pp. 13, 15 and Plate III. A corruption of the surname al-Khowārizmi has given the words 'algorithm' and 'augrim' in English and 'awgrym/awgrim' in Welsh.

44 This was promoted by Leonardo Fibonacci of Pisa (1170–1250) in his *Liber Abaci* of 1202. See L. E. Sigler, *Fibonacci's Liber Abaci: a Translation into Modern English of Leonardo Pisano's Book of Calculation* (New York, 2002); G. F. Hill, *The Development of Arabic Numerals in Europe* (Oxford, 1915).

45 *Glossary for the British Library Catalogue of Illuminated Manuscripts, www.bl.uk/catalogues/ illuminatedmanuscripts*, where it is noted that 'numeric representation . . . did not come into general use . . . until the fifteenth century'. On the introduction of Hindu–Arabic numbers in mid-fifteenth-century inscriptions, see H. T. Morley, 'Notes on Arabic numerals in medieval England', *Berkshire Archaeological Journal*, 50 (1947), 81–6; see also Peter Wardley and Pauline White, 'The Arithmeticke Project: a collaborative research study of the diffusion of Hindu-Arabic numerals', *Journal of the Family and Community Historical Research Society*, 6/1 (2003), 1–17, on late sixteenth-century increase in the use of this numerical scheme.

46 *Oxford English Dictionary* (Oxford, 1989) online edn September 2011, accessed 14.09.11, entry for *algorism*, 'Ouer the wiche degrees ther ben nowmbres of augrym', quoted from W. W. Skeat (ed.), *A Treatise on the Astrolabe addressed to his son Lowys by Geoffrey Chaucer A.D. 1391, edited from the Earliest MSS . . .*, Early English Texts Society, 16 (London, 1872), I, §7, p. 5. See also the discussion of *algorism* in David Eugene Smith, *History of Mathematics*, vol. 2 (1925, reprinted by Dover Publications, New York, 1958), pp. 9–10.

47 British Library, Sloane MS 110, f. 29r.

48 J. E. Caerwyn Williams, 'Gutun Owain', in A. O. H. Jarman, G. R. Hughes and D. Johnston (eds), *A Guide to Welsh Literature 1282–c.1550*, vol. 2 (Cardiff, 1997), pp. 240–55.

49 See pp. 133–6.

50 Dwnn, *Heraldic Visitations*, p. 68.

51 John Prise, *Yny lhyvyr hwnn y traethir. Gwyðor kymraeg. Kalandyr. Y gredo, neu bynkeu yr ffyð gatholig. Y pader, ney weði yr arglwyð. Y deng air deðyf. Saith Rinweð yr egglwys. Y kampey arveradwy ar Gwyðieu gochladwy ae keingeu* (London, 1546), 33 pp. (17 leaves).

52 R. Geraint Gruffydd, '*Yny Lhyvyr Hwnn* (1546): the earliest Welsh printed book', *Bulletin of the Board of Celtic Studies*, 23/2 (1969), 105–16.

53   For Prise's career, see ibid. See also R. Geraint Gruffydd, 'Y print yn dwyn ffrwyth i'r Cymro: *Yny Lhyvyr Hwnn*, 1546', *Y Llyfr yng Nghymru = Welsh Book Studies*, 1 (1998), 1–20. Gruffydd gives very little attention to the section on numbering, stating only that it is a section ending with 'sylw digon gogleisiol am yr "awgrym"' (rather an intriguing reference to the 'algorithm'). Prise's source is not known. It is clearly not a direct translation of Recorde.

54   See, for example, TNA series SC 6, Ministers' Accounts for the Principality of Wales, for assessments made during the later Middle Ages; and TNA series E 179 for records relating to ecclesiastical subsidies in Wales from 1291 onwards; lay assessment for subsidies in 1291 and from 1543 onwards.

55   TNA E 179/220/166 (Wrexham, first collection 1543–4).

56   34/5 Henry VIII, c. 27. For details of the grant and its collection in Wales see the E 179 database of the National Archives, *www.nationalarchives.gov.uk/e l 79.*

57   Recorde introduces Hindu–Arabic numerals in an early section on numeration in the *Ground* (sigs [A.vii]^v–C[.i]), but he continues to make occasional use of Roman numerals, alongside Hindu–Arabic numerals, either as cardinal numbers to enumerate various objects (e.g. '.ii. droues of cattell' (sig. C.ii); 'iiii. marchauntes' (sig. P.v); 'one example or .ii.' (sig. [F.viii]); 'I take 9 .iiii. tymes' (sig. G.ii')) or as ordinal numbers (e.g. 'Then come I to the .iiii. place' (sig. [E.vii]); 'and lyke wyse in the .vii. place' (sig. [E.vii]^v; 'in the .iiii. lyne' (sig. R.iii)).

58   This is a general comment, but would include the use of new forms of numbering.

59   Returns for the dioceses of Bangor, Bath and Wells, Coventry and Lichfield, Lincoln and Worcester use Roman numerals, with only St David's and the partial return for the diocese of Winchester using Arabic numerals. By contrast, extant records for the province of York use Arabic numerals. See Alan D. Dyer and David M. Palliser (eds), *The Diocesan Population Returns for 1563 and 1603* (Oxford, 2005), pp. xii, 8, 21, 43, 77, 100, 134, 144, 154, 162, 184, 258, 283. The St David's returns are in British Library, Harleian MS 595, ff. 84r–92r.

60   Andrew Petegree, 'Young, Thomas (1507–68)', *Oxford Dictionary of National Biography* (Oxford, online edn, May 2006).

61   Dyer and Palliser, *Diocesan Returns*, pp. 303, 353. See also *eidem*, pp. 328, 479 (n. 16), 502–15, 516–23.

62   E 179/223/421, dated 9 January 1544.

63   See n. 56 above.

64   James Gairdner (ed.), *Letters and Papers Foreign and Domestic: Henry VIII*, vol. 12, Part 1 (London, 1890), no. 782, p. 339.

65   H. C. Maxwell-Lyte (ed.), *Calendar of the Patent Rolls Preserved in the Public Record Office: Henry IV*, vol. 3 (London, 1907), 62, 8 July 1405, m.15d.

66   H. C. Maxwell-Lyte (ed.), *C. P. R.: Edward IV, Richard III*, 1476–1485, vol. 2 (London, 1901), 245, 27 February 1481, m.10d.

67   H. C. Maxwell-Lyte (ed.), *C. P. R.: Edward III*, vol. 1 (London, 1891), 245, 248, 6 March, 1328, 2 Edward III, Part 1, mm. 20, 22: grants of quayage and murage for seven years to aid the construction of a quay. Repeated grants of quayage were made in 1344, 1377, 1390 and 1399 in order to maintain the quay (H. C. Maxwell-Lyte (ed.), *C. P. R.: Edward III*, vol. 6 (London, 1902), 239, 24 April 1344, 18 Edward III, Part 1, m. 20; *idem* (ed.), *C. P. R.: Edward III*, vol. 16 (London, 1916), 435, 13 March 1377, 51 Edward III, m.32; *idem* (ed.), *C. P. R.: Richard II*, vol. 1 (London, 1895), 80, 12 December 1377, I Richard II, pt. 2, m. 7; *idem* (ed.), *C. P. R.: Richard II*, vol. 4 (London, 1902), 243, 5 May 1390, 13 Richard II, Part 3, m. 18; *idem* (ed.), *C. P. R.: Richard II*, vol. 6 (London, 1909), 518, 12 April 1399, 22 Richard II, Part 3, m. 40.

68   Sir Harris Nicolas, *Proceedings and Ordinances of the Privy Council of England, 1386–1542*, vol. 3: *1 Henry VI, 1422 – 7 Henry VI, 1429* (London, 1834), Chronological Catalogue, x: 2 March 1423. Freedom from payment of quayage, murage, pontage, paviage and piccage for Tenby merchants at Bristol for twelve years.

69   H. C. Maxwell-Lyte (ed.), *C. P. R.: Edward III*, vol. 16, 114–15, 8 June 1375, 49 Edward III, Part 1, m. 11; *idem* (ed.), *C. P. R.: Richard II*, vol. 1, 112, 6 February 1378, 1 Richard II, Part 3, m. 26; *idem* (ed.), *C. P. R.: Henry IV*, vol. 2 (London, 1905), 120–1, 22 August 1402, 3 Henry IV, Part 2, m. 2, where provision is made for constructing a covered shambles for the sale of meat.

70   Dwnn, *Heraldic Visitations*, p. 68.

71   See W. P. Griffith, 'Schooling and society', in J. Gwynfor Jones (ed.), *Class, Community and Culture in Tudor Wales* (Cardiff, 1989), pp. 79–119.

72   See nn. 65 and 66 above.

73   H. C. Maxwell-Lyte (ed.), *C. P. R.: Henry IV*, vol. 2, 297, 28 September 1403.

74   James Gairdner (ed.), *Letters and Papers Foreign and Domestic: Henry VIII*, vol. 12, Part 2 (London, 1891), 225–6, no. 613, 1537.

75   Gairdner (ed.), *Letters and Papers*, vol. 12, Part 1, 339, no. 782.

76   Ibid.

77   Gairdner (ed.), *Letters and Papers*, vol. 12, Part 2, 370–1, no. 1057.

78   For general accounts of Portuguese exploration, see C. R. Boxer, *The Portuguese Seaborne Empire, 1415–1825* (New York, 1970); Bailey W. Diffie and George D. Winnius, *Foundations of the Portuguese Empire 1415–1580, I: Europe and the World in the Age of Expansion* (Minneapolis, 1977); Malyn Newitt, *A History of Portuguese Overseas Expansion, 1400–1668* (London, 2005).

79   Gairdner (ed.), *Letters and Papers*, vol. 12, Part 2, 431, no. 1199.

80   See, for example, *Castle*, pp. 71–2.

81   TNA LR 2/238 Survey of burgage lands in Tenby, 1586, f. 123r. The Recordes did hold at least seven other properties in commercially advantageous places in the town in 1586. The family, however, is located by Lewis Dwnn in his 1597 visitation at 'St. John's by Tenby' (Dwnn, *Heraldic Visitations*, pp. 68–9) being the former hospice founded

by William de Valence and dedicated to St John the Baptist, but not part of the order of St John. (See John Cule, 'Some early hospitals in Wales and the Border', *National Library of Wales Journal*, 20/2 (1977), 97–130.) It is not known whether it ceased to be a hospital or hospice, becoming a Recorde residence, before Robert the mathematician was born.

82   *Whetstone*, Dedicatory letter to the Muscovy Company.

83   S. Dimmock, 'Reassessing the towns of southern Wales in the later Middle Ages', *Urban History*, 32/1 (2005), 33–45; *idem*, 'Haverfordwest: an exemplar for the study of southern Welsh towns in the later Middle Ages', *Welsh History Review*, 22/1 (2004), 1–28; *idem*, 'Urban and commercial networks in the later Middle Ages: Chepstow, Severnside and the ports of southern Wales', *Archaeologia Cambrensis*, 152 (2005 for 2003), 53–68.

84   Madge Dresser and Peter Fleming, *Bristol: Ethnic Minorities and the City 1000–2001* (London, 2007), pp. 19–48.

85   A. D. Carr, *Medieval Anglesey* (Llangefni, 1982), pp. 112–15, 240–2.

EIGHT

Commonwealth and Empire: Robert Recorde in Tudor England

HOWELL A. LLOYD

L ET ME BEGIN by noting very briefly a few key features of the condition of the England where Robert Recorde spent most of his life. During the decades from the mid-1520s to the late 1550s this England experienced political turbulence, religious change and considerable economic stress. Henry VIII, having failed to persuade Pope Clement VII to grant him a divorce from his first wife, withdrew his kingdom from obedience to Rome and embarked upon a quasi-Protestant Reformation. The process was accompanied by assertions of the doctrine known as 'imperial kingship': that 'this realm of England is an *empire* [my italics]' under 'one supreme head and king', a doctrine allegedly validated by 'sundry old authentic histories and chronicles'.[1] It was a position that Henry's immediate successors were poorly equipped to maintain. His only son, who ascended the throne at the age of nine, had barely emerged from boyhood by the time of his death. For all his manifest commitment to the Protestant cause, Edward VI amounted in one modern scholar's phrase to little more than 'an articulate puppet'[2] and never escaped from the tutelage of two successive lords protector, the rival dukes of Somerset and Northumberland. Henry's elder daughter, child of that first discarded wife, seemed intent during her five-year reign upon vindicating her Spanish mother as she strove to make England obedient again to Rome and herself pleasing to a Spanish royal husband.

All three reigns saw recurrent eruptions of violence in the forms of offensive and defensive wars abroad, and, at home, threats and actual outbreaks of rebellion. In most instances the rebellions involved economic problems to varying degrees. In the course of Recorde's thirty or so English years, prices of consumables rose by approximately 90 per cent with grain price inflation especially pronounced, while agricultural workers' purchasing power fell by some 26 per cent. As for commerce, English foreign trade, which was largely in woollen cloth, experienced a relative boom period in the first half of the sixteenth century. Currency debasements under both Henry VIII and Edward VI rendered English products cheaper in foreign markets. But the profitable trading activity stimulated efforts to increase wool production, the effects of which threatened 'utter dissolation to the whole Realme, furnished only with shepe and shepherdes in stead of good men'.[3] Thus the author of the *Discourse of the Common Weal*, drafted in 1549, bemoaned the condition of English society in terms which echoed those of Raphael Hythlodaye in Thomas More's *Utopia* (1516): 'a certain conspiracy of riche men procuringe theire owne commodities under the name and title of the *commen wealth*' (my italics).[4]

All these features of England's mid-Tudor history have a bearing upon Recorde's career and his writings. A further feature warranting emphasis here is the intellectual and educational context of his work. The principal driver of our period's intellectual history is generally described as 'humanism', a phenomenon characterized chiefly by the rediscovery and reinterpretation of classical Greek and Roman texts and assimilation of the values represented in them. In pursuing their objectives the humanists are held to have developed new approaches, new perspectives, new critical methods. Predominantly literary and linguistic, their concerns and their influence extended, omnivorously, to virtually all branches of knowledge, from theology and legal institutions to mathematics, astronomy and the remains of the past, and to the teaching of these and other subjects, reinvestigated and even reorganized for the purpose. Thus, by his death in 1476 the German scholar Johannes Müller (alias Regiomontanus), protégé of the Greek scholar and humanist patron Cardinal Joannis (Basil) Bessarion, had come close to completing the translation and commentary on Ptolemy's *Almagest* begun by his teacher Georg Peurbach, himself the author of *Theoricae novae planetarum* (1472), which constituted an improvement on the existing literature in the field. German mathematicians – Widman, Riese, Rudolff, Apian, Stifel – had, by the mid-1540s, made contributions in the field of algebra culminating in that 'tremendous stimulus to algebraic research', the *Ars magna* (1545) of the Italian humanist Girolamo Cardano, physician, mathematician and alleged heretic.[5] Researchers in mathematics, notably but not exclusively at Padua, brought fresh arguments to bear upon the vexed topic of the epistemological significance of mathematical demonstration, the relation between the objects of mathematics and those of the material world, as expounded by that now much-criticized but still most influential of all philosophers, Aristotle.[6] As for teaching, by Cardano's time a number of continental universities were seizing avidly upon opportunities for innovation in subject areas and modes of study. Thus, at Wittenberg, where the great scholar and Reformation theologian Philip Melanchthon taught, lectureships had been instituted by the mid-1540s in Greek, Hebrew and physics as well as two in mathematics, and programmes were radically revised in other areas, including logic and ethics as developed by Melanchthon himself.[7] But the state of educational affairs at England's ancient universities was markedly less exciting – though perhaps not as moribund as is sometimes supposed.

Over the preceding two centuries Oxford, where Recorde graduated BA in 1531, had been at the forefront of European scholarship in mathematics and astronomy – two intimately related fields, both closely linked with medicine. Those days had passed, owing possibly to the weakening of academic links with the leading continental institutions, especially the University of Paris, during the decades of the Black Death and the Hundred Years War.[8] If we rely on the evidence of the University statutes, these seem to suggest that the student curriculum stagnated and that, in Recorde's time, it continued to be governed largely by the medieval regulations.[9] However, the evidence is insufficient. A litmus test of where educational institutions stood in relation to the initiatives of Renaissance humanism may be found in how far they adopted the study of Greek, which was certainly available at Oxford. William Grocyn was instrumental in introducing it at Magdalen College where he was reader in divinity from 1481. He was supported by Thomas Linacre, who accompanied Grocyn to Italy in the late 1480s specifically to study Greek and also, in Linacre's case, medicine at Padua, that powerhouse of humanism in an academic setting where half a century later Cardano was to teach. Combining these interests, Linacre produced new translations of the works of Claudius Galen directly from the Greek, and so vastly improved understanding of what that great authority had actually said.[10] Linacre was later one of the founders of England's College of Physicians (1518), which had close links with Corpus Christi College, itself founded in the previous year by Bishop Richard Fox with provision for the study of Greek specified in the college statutes. In 1523 Cardinal Wolsey pronounced in the regulations for his new Cardinal College that Greek should be taught there every day at 1.00 p.m. Having dispensed with Wolsey and transformed his college into Christ Church, Henry VIII, himself no mean classical scholar, appointed in 1540 five new Regius professorships at both Oxford and Cambridge, in Theology, Civil Law, Hebrew, Medicine and, of course, Greek once more.[11]

While neither all nor any of this makes Oxford an exemplar of progressive education in Renaissance Europe, it does indicate that the student experience of men such as Recorde is not to be judged simply on the basis of what the university statutes prescribed. What especially mattered, then as now, was the tuition available in the colleges. We do not know of which college Robert Recorde became a member, but if it were Magdalen or, better still, Corpus, that would go some way towards explaining the pride he took in his knowledge of Greek. As for medicine, it is certainly intriguing that, while Linacre (d. 1524) made provision in his will for two lectureships to be founded in that discipline at Oxford and one at Cambridge, Recorde decided to leave the former university for the latter to pursue his medical studies. Upon closer examination, this need not surprise. At Oxford, difficulties amounting to outright opposition delayed implementation of Linacre's plans. In contrast, Cambridge, where the arts undergraduate curriculum had been reformed in 1488 in favour of 'humane letters', swung into action right away.[12] The Cambridge medical lectureship was established at St John's and its first holder appointed in 1525, the year immediately following Linacre's death. At precisely the same time St John's adopted new statutes partly – though by no means wholly – based on Richard Fox's at Corpus Christi, Oxford. They included provision for delivery of lectures on mathematics in the long vacation by four men instead of the previous one, and for the students to be 'exercised' by four examiners on what they had learned from lectures in philosophy, logic, 'studies which are called humanities', and mathematics. At St John's once more, praelectors were instituted in Greek and Hebrew, on the strength of funds provided by one of the college's great benefactors, Lady Margaret Beaufort, great-granddaughter of John of Gaunt, widow of Edmund Tudor and mother of Henry VII.[13] In company with Christ's, St John's was a leading influence upon curriculum development in early sixteenth-century Cambridge, and an important factor in that university's being regarded as more in the van of educational innovation than Oxford at that time.

Of course, even in Cambridge the disposition to innovate was firmly circumscribed. Greek might be available and delivery mechanisms modified, but the received framework of knowledge remained firmly in place. That framework was, and remained, essentially Aristotelian. On the Continent revolutionaries such as Luther might attack the very foundations of Aristotelian learning and dismiss as altogether without value such prime Aristotelian philosophical preoccupations as 'matter, form, measurement and time'.[14] In the very year that Recorde's *The Ground of Artes* first appeared in print (1543), the self-advertisingly radical Parisian teacher Pierre de La Ramée was publishing his scandalous 'Aristotelian censures' (*Aristotelicae animadversiones*) and 'Divisions of Dialectic' (*Dialecticae partitiones*), works which announced what has been termed 'an anti-Aristotelian programme'.[15] No comparable critical attack upon the philosopher nor upon anyone else seems to have figured explicitly in Recorde's student experience, nor indeed in his own thought thereafter. On the contrary, there are grounds for holding that, despite appearances, his scientific thinking continued to be grounded most firmly upon the materials that had served his medieval predecessors.

The Vrinal of Physick opens, boldly enough, with a warning against 'vayne and disceytefull' books, (sig. [A.vi]ᵛ) coupled with a splendidly ecumenical recommendation of the works of Pope Julius II's physician, Giovanni da Vigo (1450–1525), as translated by Recorde's exact contemporary and fellow student at Oxford, Bartholomew Traheron (*c*.1510–*c*.1558), later a Protestant controversialist. There follows a resounding claim that the *Vrinal* 'is written according to the myndes of the most excellent writers of Physicke, bothe of the Grekes and of the Latins' (sig. B.ii). But these, promptly listed, turn out to be a motley crew. Their writings range from such long-established standard works as the *De materia medica* (first century AD) of Dioscorides Pedanius of Anazarbus (in modern Turkey) and the eleventh-century Arab philosopher Avicenna's *Libri canonis*, to the compilations of various Byzantine physicians of uncertain identity[16] or dubious repute, together with Quintus Serenus Sammonicus' *Liber medicinalis* (third century AD), a verse composition based largely on Pliny the Elder's *Natural History*. In the event, most of these figure only incidentally, if at all, in the *Vrinal*'s subsequent pages where the bulk of the material is traceable to Hippocrates and Galen.

In *The Castle of Knowledge*, the Scholar twice singles out three writers on whom he has elected to concentrate his attention. Two of them, the fifth-century Neoplatonist Proclus and the thirteenth-century mathematician John of Sacrobosco (or Holywood), were mainstays of the medieval curriculum; the third, 'Orontius the Frenche man' (*Castle*, p. 98), was Recorde's contemporary, the versatile mathematician, astronomer and creative cartographer Oronce Fine (1494–1555).[17] The Master's recommendations in response are as eclectic as ever, drawing on Euclid's *Phaenomena*, the work of the Greek Cleomedes, and that of a clutch of medieval Englishmen (Robert Grosseteste, Michael Scott, William Batecombe, John Baconthorpe) followed by Pliny and others for more advanced study. His warmest recommendation is reserved for the commentary on Proclus by Joannes Stöffler (1452–1531, professor of mathematics at Tübingen), 'whyche booke I wishe were well recognised (as it hathe great neede) then myghte it serue in steede of a greate numbre of other bookes' (*Castle*, p. 98). Yet in the body of the work it is directly from Proclus and Sacrobosco that he cites by far most frequently.

But it would be utterly wrong to dismiss Robert Recorde as merely a purveyor of conventional knowledge in the vulgar tongue decorated with a certain amount of contemporary reference. The grounds are ample for identifying him with the humanist approach to learning. Chief among them is his pride in his mastery of Greek and his eagerness to deploy it for critical purposes. Errors on the part of earlier scholars might be excusable in view of 'the ignorance of that time, for lack of knowledge in the Greeke tonge'. Indeed, 'Ignorance of the Greeke tongue hathe hindred muche manye good wittes: whiche maye often appeare not only in

good Iohn de Sacro bosco, but also in many writers within these 300. yeares espe-
ciallye.' Now, however, there was no excuse: thus, 'Cleomedes the greeke authour
is very woorthye to bee often readde: but beste in hys owne tongue, for the latine
booke is muche corrupted' (*Castle* pp. 178 [171], 195, 98).

Yet Recorde's use of the critical approach was neither confined to considerations of language nor conducted with delicacy of tone. In *The Whetstone of Witte*
the elucidations offered by others of 'the Rule of Algeber' or the 'rule of equation' were dismissed as 'idle bablyng'. The one exception was Johann Scheubel
(1494–1570), another of the Tübingen mathematics professors, translator of
Euclid and author of various mathematical texts, including a 'Concise and easy
account of algebra' (*Algebrae compendiosa facilisque descriptio* (1552)). Neither concise
nor easy enough for Recorde's taste, Scheubel's account of 'Algeber's rule' was
discussed at length in the *Whetstone* and pronounced overelaborate, Recorde providing his own exposition with fewer distinctions and more illustrative examples
(*Whetstone*, sigs Ff.i–Gg.i^v). Such discussions tallied precisely with his repeated
warnings against unquestioning acceptance of received opinion, no matter how
eminent the source. When at the very opening of *The Ground of Artes* the Scholar
declares to the Master his willingness 'to consente to your sayenge and to receaue
it as truth: though I se none other reason, that doth leade me thervnto', he is
immediately charged with 'blynde ignoraunce' (sig. A[.i]). 'Take heed', comes one
of several subsequent warnings, 'in al mennes workes, [that] you be not abused
by their autoritye, but euermore attend to their reasons, and examine them well'
(*Castle*, p. 127 [129]). On the face of it, the adjuration tallied well enough with
Aristotle's pronouncement in the *Posterior Analytics*, a text on which Recorde may
possibly have lectured at Oxford whilst preparing for his master's degree: 'when
there is demonstration, a man who has not got an account of the reason why does
not have understanding.'[18] But Recorde was no servant of convention, however
well entrenched. Indeed, in his view of how mathematical knowledge should be
organized and presented for educational purposes he anticipated one of the more
unorthodox thinkers of the age.

Nine years after Recorde's death Pierre de La Ramée, better-known as Ramus,
delivered to the French Privy Council in the Royal Chamber at the Louvre, in the
presence of King Charles IX, a characteristically polemical *Remonstrance . . . concerning the Regius professorship in mathematics*. The key passage might have been composed
by Recorde himself:

Order in mathematics in no way resembles history where one can examine a
passage at the end, in the middle, at the beginning without taking account
of what goes before. In mathematics order is not only profitable and useful, but absolutely necessary. The first of these [mathematical] disciplines is

arithmetic, the art of numbering everything that can fall within the scope of number, adding, subtracting, multiplying, dividing all whole and fractional numbers when comparing their *raisons* [= 'ratios' – or, arguably, 'grounds'] and proportions. The second is geometry, the art of measuring everything subject to measure, such as length, breadth, height and in general all sizes of planes and solids, whether in heaven or on earth, or of some other measurable subject. This second discipline can in no way be understood nor practised without the first: for to measure is to number intervals, it is to compare the *raisons* and proportions of figures. Astrology, which follows, can likewise not be conceived nor demonstrated without arithmetic and geometry: for astrology is nothing other than arithmetic [applied] to numbering the degrees, minutes and all other parts and movements of the celestial bodies; it is nothing other than geometry for measuring the triangles, circles, spheres and all figures existing there. And so the other mathematical disciplines, even beyond the propositions of arithmetic, geometry, astrology, are constructed in such order that not to know the first is not to understand the second, and not to know the first and the second is not to understand the third. In short, if a scholar has missed a single lesson in mathematics, he needn't trouble to return to school, for he will understand nothing that follows … Whoever embarks upon mathematics against the order of mathematics betrays his inadequacy and his ignorance.[19]

This passage, and the *Remonstrance* itself, was in fact part of a two-pronged campaign on Ramus' part. His first, and abiding, objective was widespread among humanists: to simplify and bring order to pedagogic methods. His second and immediate aim was to be appointed in place of a rival to the Regius professorship. We know of no comparable ambition in Recorde's case: indeed, unlike Ramus, Recorde was singularly modest in his estimation of his own professional deserts. Even so, he was certainly a subscriber to the humanist pedagogic programme and, in respect of the sequential teaching of mathematics, the famously acerbic Ramus could hardly have offered a neater outline of the route which Recorde had already constructed, from *Ground* to *Castle* via the *Pathway* of mathematical knowledge.

It is evident, then, that Recorde evinced many of the defining characteristics of the humanist intellectual: the linguistic concern, the critical approach, the readiness to reorganize knowledge and to revise its presentation for teaching purposes, even his adoption of the dialogue form as a pedagogical tool.[20] Significant also in this connection is the range of his interests. Some indication of the contents of his library may be gleaned from the 'Index of British writers' (*Index Britanniae scriptorum*) compiled by the historian, polemicist and Irish bishop John

Bale (1495–1563). Intended to offset the 'moste horrible infamy' of the destruction of the monastic libraries under Henry VIII, the *Index* continued the work of the antiquary John Leland by listing books, which Bale himself had either seen in various repositories or had had drawn to his attention by friends and acquaintances. Among them was Recorde, whom Bale credited in his *Index* as his source for some forty-five titles. Only two of these were specifically texts on mathematics, and only one on medicine. In addition to ten astronomical works, there were seven in the field of antiquities, half a dozen on law, nine in the field of theology, and an assortment of works ranging from the prophecies of two Merlins of Celtic mythology (Ambrosius and Sylvester) to the elementary Greek grammar commissioned by King Henry from David Talley, physician and (according to Bale) 'English hammer of the papists'.[21] While these titles by no means constituted the whole of Recorde's collection of books and manuscripts,[22] they represent well enough the omnivorousness of his interests and the eclectic nature of his resultant collection. And there is more than this to be gleaned from the Bale connection.

The *Index* notes over one hundred individuals from whom its compiler had gleaned items of bibliographical information. Bale, evidently, was acquainted with them all. Of course, it is utterly improbable that the circle of his acquaintances was complete in the sense that everyone in it knew everyone else. Equally, it can scarcely be supposed that the members of Bale's circle knew no one in it other than him. There is clear evidence that at least two of them were known to Recorde, who was among Bale's most useful suppliers of material. Robert Talbot, antiquary in his own right and source of twenty-two titles for the *Index*, lent Recorde two specific items from his collection of Anglo-Saxon manuscripts. The printer Reyner Wolfe, native of Gelderland and publisher for no less a personage than Archbishop Thomas Cranmer, printed the first editions of the first three of Recorde's mathematical textbooks. Several among the many printers and booksellers who advised Bale had Welsh connections. Robert Crowley, sometime Fellow of Magdalen College, published works by Recorde's compatriot William Salesbury, translator of biblical and liturgical texts into Welsh and of Linacre's Latin version of Proclus' *De sphaera* into English, and himself a contributor to Bale's enterprise. Edward Whitchurch, publisher of the Great Bible under the aegis of Thomas Cromwell, also published the earliest book printed in Welsh, the collection of religious texts titled *Yny Lhyvyr Hwnn* (1546) and attributed to Sir John Prise, an important informant of Bale's and an energetic collector of manuscripts from the monasteries, which, as an agent of Thomas Cromwell's, he helped to close.[23] Yet another of Bale's informants, Thomas Huet, was to collaborate with Salesbury and Bishop Richard Davies in translating the New Testament into Welsh. Of course, Recorde himself may not have been personally acquainted with any of his fellow Welshmen. But factors other than simple personal acquaint-anceship bound together the members of Bale's circle. Most of them were, like Bale himself, adherents of the Protestant version of the Christian faith, often in its evangelical guise. And, apart from religious orientation, a great many of them were demonstrably enthusiasts for humanist learning on a grander scale than merely a collector's fondness for *scriptores Britannici*. Some were themselves active humanist scholars, among them such distinguished figures as John Cheke, student and Fellow of St John's College, Cambridge's first Regius professor of Greek and tutor to the future Edward VI, at whose court he emerged as a leading personal-ity. Some were patrons of humanist learning, including noblemen of the standing of Charles Blount, fifth Baron Mountjoy, dedicatee of the *Adages* of Desiderius Erasmus himself, and of the distinguished educationist Juan Luis Vives's 'On the ground of youthful studies' (*De ratione studii puerilis*). And some were school-masters: among them William Horman, headmaster of Eton and then of Winchester, and author of several school textbooks on Latin grammar; and Ralph Radcliffe, who established his own school at Hitchin and had the temerity to challenge Cheke's authority on Greek pronunciation.

So, in so far as he belonged to Bale's circle, Recorde was linked in terms at least of common interests on the one hand with proponents of religious change and on the other with supporters of humanism and educational reform. Such support was confined neither to the associates of John Bale nor to the humanists' favourite subject area of linguistic and literary studies. It emanated also from such political operators as Richard Whalley, another product of St John's College, subsequently chamberlain and man of business for the Duke of Somerset, and, by Recorde's own account, 'one of them that both loued & also moch desyreth to further good learnyng' (*Ground*, Preface, p. [ii]). It emanated too from the example of King Edward VI himself, whose 'louyng subiectes' could 'see in your highnes not only suche towardnes, but also suche knowledge of dyuers artes, as seldome hath bene sene in any prince of such yeres' (*Ground*, 1552, Preface to 'the Kynges maiestie', sig. [A.viij]). The words are Recorde's but, for his particular purposes, the support in question resided most encouragingly of all with his fellow cultiva-tors of medicine and mathematics. In *The Vrinal of Physick* he applauded 'that wor-thye knight & lerned clerke, syr Thomas Elyot' (sig. [B.vii]), distinguished pupil of Linacre, author of *The Castel of Helth* (1539?), which he dedicated to Thomas Cromwell and, in his own best-known work, advocate of 'abundant salaries' for schoolmasters so that they might then 'induce their herers to excellent lernynge'.[24] And as for mathematics, one of Tudor England's at once most celebrated and most notorious practitioners of that art was yet another of Recorde's compatri-ots, at least by descent, and another product of Cheke's Cambridge college of St John. This was John Dee, who described himself as Recorde's 'friend', edited later editions of *The Ground of Artes*, and composed a *Mathematicall Praeface* for Henry Billingsley's translation of Euclid's *Elements of Geometrie*, which appeared in 1570 under the imprint of another of Bale's contributors, the evangelical printer and bookseller John Day. In his *Praeface* Dee famously waxed rhapsodic about the potential of studying mathematics:

> By *Numbers* propertie therefore, of us, by all possible meanes, (to the perfec-
> tion of the Science) learned, we may both winde and draw our selues into
> the inward and deepe search and vew, of all creatures distinct vertues, natures,
> properties, and *Formes*: And also, farder, arise, clime, ascend, and mount vp
> (with Speculatiue winges) in spirit, to behold in the Glas of Creation, the
> *Forme* of *Formes*, the *Exemplar Number* of all thinges *Numerable*: both visible and
> inuisible: mortall and immortall, Corporall and Spirituall.[25]

Here was an extreme version of a vision which other mathematical practitioners shared without indulging in Dee's metaphysical effusions. Recorde himself,
who, Neoplatonist-fashion, saw mathematics as affording the key to understanding of 'The Sphere of Destinye' (*Castle*, title page) and even to 'the knowledge
of God, and his highe mysteries' (*Castle*, p. 284), kept his feet on the ground
much more firmly than Dee when compiling his own manuals of instruction. But
what numerous mathematical practitioners did share, with Dee, with Recorde
and with a great many others was an explicit concern for the condition of what
they termed the 'commonwealth'. This term, which has provoked a good deal of
comment and debate amongst historians, warrants our dwelling briefly upon it in
order to enhance a little further our appreciation of Recorde's motivation and his
contribution in the context of the society of his time.

In sixteenth-century usage the English term 'commonwealth' has a range of
connotations. They are rooted in a combination of the classical '*res publica*' as
discussed in particular by Cicero, and the medieval concept of '*bonum commune*',
the 'common good'. '*Res publica*' relates essentially to the political: in Cicero it
can signify the form as well as the business of government, and can also refer to
the political collectivity (or people), though by no means necessarily implying
a non-monarchical or acephalous political order – a 'republic', in modern par-
lance. '*Bonum commune*' has to do essentially with the social: the good of society,
which encompasses the well-being of the individual as well as the community
at large, promoted and safeguarded by justice or good laws, and embracing the
materially advantageous (*communis utilitas*) as well as the virtuous or morally com-
mendable. All of these elements fed into the concept signified by the terms 'common weal' and 'commonwealth' which seem to have become current in England
from about the middle of the fifteenth century.[26] They were terms that not only

accommodated the idea of the well-being – good order, security, prosperity – of the community at large (*bonum commune*), but also served as vernacular equivalents of *res publica* with all the political and ethical significance that good classical scholars would associate with that term.

Scholars are still investigating the evolution of the terms we have just reviewed, and how their development relates to contemporary political and social developments. Critical comment has been especially sharp in relation to a phenomenon once identified as the 'commonwealth party'. That sometime doyen of Tudor historians, A. F. Pollard, claimed over a century ago to have discerned such a body in the England of Edward VI;[27] and seventy years later it was identified afresh by Whitney R. D. Jones as a group desirous that secular political authority be brought to bear specifically upon the social and economic problems of the time.[28] Notwithstanding ill-founded conjectures and scholarly scepticism concerning 'party',[29] there is abundant evidence of the presence of 'commonwealth' in contemporary discourse not merely as a 'catchword',[30] but as a vehicle for a range of propositions that embraced all the time-honoured connotations of the term, and more. Thus, for the humanist Thomas Starkey, product of Magdalen College, Oxford, a 'true commyn wele' was 'no thyng els but the prosperouse & most perfayt state of a multytud assemblyd togyddur in any cuntrey, cyty or towne, governyd vertusely in cyvyle lyfe accordyng to the nature & dygnyte of man'.[31] Here indeed was a Ciceronian *res publica*. To it John Hooper, bishop of Gloucester, added an evangelical's assessment of how to promote a commonwealth's moral and material well-being: that its people obey God's will and observe the established order, that they should 'know eche of them there dewtes', that there should be 'lawes to preserue' their 'persones' and laws also 'to preserue souche goddess as appertayne unto' them.[32]

Such considerations figured afresh in a prescription for a 'well-ordered commonwealth' penned by none other than Edward VI himself for remedying the social and economic problems of his realm, afflicted as it was by 'sores' that must 'be cured with these medicines or plasters'. The remedy which the young king ranked first in 'order' in 'dignity and degree' was 'good education', a proposition which the royal product of humanist instruction hammered home with a quotation from Horace: *Quo [semel] est imbuta recens servabit odorem testa diu* – in Edward's own translation, 'With whatsoever thing the new vessel is imbued it will long keep the savour'.[33] It was a position which his kingdom's mathematical practitioners enthusiastically endorsed.

'Fauour me', wrote Leonard Digges in his *Pantometria*, a work on practical mathematics, posthumously published, 'as I render the furtheraunce of good learninges, profitable to a common wealth'.[34] 'My onely vocation', declared the schoolmaster John Mellis, editor and amplifier of Recorde's *The Ground of Artes* and in his own right author of a work on book-keeping, 'hath bene (thinking it a meete exercise for a common welth) in training vp of youth to write and draw, with teaching of them the infallable principles and briefe practises of this worthie Science.'[35] Affirming the utility of the textbook on geometric rules for accurate land measurement compiled by his 'frynde' and colleague Richard Benese, the translator and former Augustinian canon, Thomas Paynell, observed in his preface to that work:

> There are lykewyse, that after longe disputation do not onely assygne euery science hys peculyare laude and prayse, but also that discusse whether of such noble sciences are moost for the commune weale supposynge (and well) that thynge to be mooste excellent that is mooste for euery mans profyte: and that commune utylite and profyte dothe none other wyse exceed priuate gayne and profyte, than golde all other metalles.[36]

Another translator, Richard Eden, sometime employee of Richard Whalley, stated the point more succinctly: 'in certeyne small and obscure members of the common wealth consisteth no small increase to the perfection of the whole.'[37] But none put the case more plainly than Recorde himself, echoing Cicero while affirming the merits of his own discipline in the preface to the 1552 edition of *The Ground of Artes* which he addressed to none other than King Edward: 'informed reason was the onely instrument, or at leaste the chiefe meane to brynge men vnto ciuile regiment, from barbarous maners & beastly conditions,' the 'reason' in question being specifically mathematical, indispensable to such essential affairs as 'iuste partition of landes', all kinds of 'byinge and sellynge', and even maintaining 'the trewe orders of Justice' and

> degeres of estates in the commune wealthe . . . Wherfore I may wel say, that seing Arithmetike is so many wais nedefull vnto the fyrst plantyng of a comon welthe, it must nedes be as muche required to the preseruation of it also; for by the same meanes is any common wealthe continued, by whyche it was erected and establyshed. (*Ground*, 1552, sigs A.iiij, [A.vj]ᵛ)

So it was by virtue of a training in mathematics that the well-being of the people, severally and collectively, might best be served, and thereby that of their commonwealth too. But one celebrated mathematician of Recorde's close acquaintance carried the argument farther. Thanks to the Billingsley translation of Euclid, declared John Dee, and to his own *Mathematicall Praeface* to that work, 'The Universities, the Storehouses & Threasory of all Sciences and all Artes, necessary for the best and most noble State of Common Wealthes' would be 'the more regarded, esteemed and referred unto'. And mathematics furnished the key to much more than the reputation of universities and the well-being of the commonwealth as traditionally construed. In his *General and rare memorials*, a work published by John Day, Dee declared himself to be offering '(almost) a Mathematicall demonstration, next under the Mercifull and Mighty Protection of God, for a faesable Policy, to bring or preserue this Victorious Brytish Monarchy, in a maruellous Security: Whereupon, the Reuenue of the Crown of England, and Wealth-Publik, will wonderfully encrease and florish.'[38] 'Security' signified on the one hand domestic peace and good order through improving the material circumstances of the people at large – the marks of a sound 'commonwealth', as the mathematicians had recognized. On the other hand there was external security, ensured through protecting the realm from foreign enemies and enhancing its power. That necessary power Dee diagnosed as maritime above all else. It should be actualized through investment in a 'Pety-Navy-Royall' and command of the 'Perfect Art of Navigation', the fruit of his own 'great knowledge in the Sciences Mathematicall and Arts Mechanicall'. Navy and navigation together constituted 'the onely Maister key, wherewith to open all Locks, that kepe out or hinder this Incomparable Brytish Impire'.[39]

Although Dee has regularly been held to have invented the term 'British Empire', there are grounds for holding that in deploying that specific term he was anticipated, and by none other than the antiquary and map-maker Humphrey Llwyd of Denbigh.[40] Others had also anticipated him, to a limited degree. The term 'empire' occurred, famously, in Henry VIII's Act in Restraint of Appeals to Rome with which we began.[41] In vouching it to warrant the realm's withdrawal from papal jurisdiction, the king's advisers drew upon a key source among the 'sundry old authentic histories and chronicles' to which the statute referred. The source figured in an assortment of evidence, 'The sufficiently copious collections' (*Collectanea satis copiosa*) which purported to demonstrate the legal and historical validity of ascribing supreme executive authority (*imperium*) not simply to the king, but to the kingdom itself. It adduced information traceable in large measure to Geoffrey of Monmouth's *History of the kings of Britain* (*c*.1139)[42] with its copious enough account of the deeds of King Arthur and substantial schedule of the prophecies of that composite figure, Merlin, most of them cheerfully invented by Geoffrey himself. Drawing heavily upon this source, the *Collectanea* noted a number of territories and islands 'which in law pertain and undoubtedly belong to the crown and office of the kingdom of Britain (*quae de jure spectant et sine dubio pertinent coronae et dignitati regni Britanniae*)', such that 'in law that excellent and most illustrious crown can and should be called rather an empire than a kingdom (*de jure potius appellari potest et debet excellentia illustrissime predicte corone imperium quam regnum*)'.[43] Now in the sixteenth century it was incontrovertible that the particular

157

territories so identified – Essex, Mercia and the Danelaw – did indeed form, and had long formed, part of the English realm. If consolidation of contiguous petty kingdoms into a political unit sufficed to constitute an 'empire', then the evidence in question suited Henry VIII's imperial purposes very well.

Yet Geoffrey of Monmouth's *History* furnished John Dee with grounds for a much more expansive view of the 'British empire'. For it described how that most famous of the kings of Britain, King Arthur, had not only thought of subduing the whole of Europe (*totam europam sibi subdere*), but had in fact gone far towards doing so, beginning with Ireland, Iceland and 'Gothland' (southern Sweden), and continuing until 'there remained no prince of any importance this side of Spain (*non remansit princeps alicuius precii citra hyspaniam*)' who did not, when summoned, come to pay him homage.[44] By Dee's own admission, his arguments on grounds of historical precedent for the re-creation of a 'British Empire' in northern latitudes 'depende cheiflie upon our kinge Arthur'.[45] Another, equally Welsh, precedent underpinned his subsequent case for 'British' expansion into the Americas: the myth of 'the Lord Madoc, sonne to Owen Gwynedd, Prince of Northwales, [who] led a Colonie and inhabited in Terra Florida or thereabowts'.[46] But there is no evidence that the Madoc legend figured in the thinking of Robert Recorde, to whom it is long since time for us to return.

Whether or not he knew of Madoc, Recorde was certainly familiar with the Arthurian legend and its message concerning imperial expansion into northern latitudes. Among his books, as recorded by Bale, were numerous vaticinatory works, including several versions of the 'prophecies of Merlin'; and Geoffrey of Monmouth was also, it would seem, the source of information which Recorde himself was credited with adding to the 1559 edition of the *Chronicle* compiled by one of London's former civic leaders, Robert Fabyan (d. 1513). The addition was the names of five British kings, who were described as 'found in certaine old petigres',[47] but were in fact the legendary perpetrators of the civil war which followed the death of Gorboduc (Gwrvyw) as recounted in the *History of the Kings of Britain*.[48] Prophecy, legend and speculative history apart, Recorde's interest in ancient and especially Anglo-Saxon chronicles embraced King Alfred's account of the voyage (*c*.880) by the Viking Ottar [alias Othere; also Ohthere] into the White Sea.[49] He believed his *The Castle of Knowledge* 'all readye' to have 'giuen some lighte' in the venture conducted 'nowe of late' on behalf of 'that woorthye companye of our Englishe marchaunts for Moscouia' (*Castle*, p. 188).[50] The same work was among the books which Martin Frobisher took on his 1576 voyage in search of a North-West Passage to Cathay.[51] Frobisher later received instruction from Dee on navigation,[52] the subject on which Recorde promised the Muscovy Company to produce 'soche a booke of Nauigation, as I dare saie, shall partly satisfie and contente, not onely your expectation, but also the desire of a great nomber beside' (*Whetstone*, sig. a.iii). Other maritime venturers consulted him: there is evidence, for instance, that a certain merchant named Philip Jones drew encouragement from 'Doctor Recordes conference in my house' about the feasibility of a North-West Passage, and also recorded how a 'plott of the West India … doth agree with the opinion of Doctor Recorde' and 'Mr Bastian Cabotta'.[53] The latter, Sebastian Cabot, to whose circle other mathematicians such as Richard Eden also belonged, was the first governor of the Muscovy Company, an organization that thought well enough of Recorde to sponsor public lectures by him in London.[54]

So Recorde with his expertise and his publications was a man of some significance in the eyes of promoters of the maritime expansion upon which Tudor England was embarking, a process that led ultimately to the creation of an overseas 'empire' in the modern sense, albeit not quite in the form envisioned by John Dee. It was a process for which, as Dee and Recorde argued and the venturers themselves evidently accepted, the support of mathematics was indispensable. But Recorde saw himself not only as supporting the ventures of others, but as engaged upon an exploratory enterprise of his own. In following him, his readers, he warned, were undertaking to tread 'straung paths' which 'muste needes be comberous, wher none hathe gone before'; they must expect 'staggeringe and stomblinge, and vnconstaunt turmoilinge: often offending, and seldome amending' (*Pathway*, p. 1, 'To the gentle reader'). Here was exploration indeed – an intellectual voyage of discovery for all who would undertake it, however knowledgeable the master himself about the course to be plotted and the destination which lay ahead. *The Ground of Artes* would lead to *The Pathway to Knowledg*, thence to *The Castle of Knowledge* and ultimately to *The Treasure of Knowledge* within. Sadly, the *Treasure* as such proved in the event to be no more a reality than the inflow of North American gold that Frobisher's sponsors were led to believe would result from his transatlantic voyages.

Recorde, then, in common with so many Tudor venturers by sea, failed finally to deliver what he had promised. Upon his activities in England, however, it seems worth bringing one final perspective to bear. The view exists that, as the process of geographical discovery and empire-building unfolded, a comparable development was beginning to occur in the sphere of mathematics itself. Formerly approached as 'the epitome of disembodied reasoning', an affair of 'unfaltering deduction from general postulates' treated as certain, mathematics, at least to some practitioners in 'early modern' England and elsewhere, was becoming less 'concerned with the elaboration of universal truths' than 'with the exploration of unknown objects' – 'an adventurous journey, a voyage of exploration and discovery in search of hidden marvels and gems'.[55] We are told that such an attitude, best exemplified in the work on infinitesimals culminating in the findings of Leibniz and Newton in terms of the calculus, was already apparent in the approach of earlier practitioners such as Thomas Harriot, whose use of Recorde's 'equals' sign stimulated its wider adoption, and who was himself the first significant mathematician to tread North American soil.[56] It would be good to end this chapter by discovering grounds for regarding Recorde, that adviser of would-be discoverers, that conductor of explorations by his own account, as in some degree an early exponent of what might be termed the 'empirical' approach to mathematical inquiry. The evidence, however, is to the contrary. Although Recorde readily affirmed that 'neither is there certaintie in any thyng without [number]', he had no doubt that mathematics did indeed yield certain knowledge, 'approued truthes'.[57] That capacity resided in mathematics, and in mathematics alone. 'It is confessed amongeste all men, that knowe what learnyng meaneth', he pronounced, 'that beside the Mathematicalle artes, there is noe vnfallible knoweledge, excepte it bee borowed of them' (*Whetstone*, sigs b.i, b.i^v). Mathematics did supply such knowledge; of this he had no doubt. It was a position, grounded in a Pythagorean view of the cosmos, which subsequent generations of philosophers and mathematicians were to find insufficient.

Notes

1   24 Henry VIII, c. 12: Act in Restraint of Appeals to Rome.

2   Dale Hoak, 'Rehabilitating the Duke of Northumberland: politics and political control', in J. Loach and R. Tittler (eds), *The Mid-Tudor Polity, c.1540–1560* (London, 1980), p. 43.

3   Anon., *A Discourse of the Common Weal of this Realm of England*, ed. E. Lamond (Cambridge, 1929), p. 52.

4   Sir Thomas More, *A Fruteful and Pleasant Worke of the Beste State of a Publique Weale, and of the New Yle called Vtopia*, trans. Raphe Robynson (London, 1551), sig. S1.

5   C. J. Boyer, *A History of Mathematics* (New York, 1991), pp. 273, 287.

6   N. Jardine, 'Epistemology of the sciences', in C. B. Schmitt and Q. Skinner (eds), *The Cambridge History of Renaissance Philosophy* (Cambridge, 1988), pp. 694–5.

7   See S. Kusukawa, *The Transformation of Natural Philosophy: The Case of Philip Melanchthon* (Cambridge, 1995), notably pp. 34, 179.

8   J. D. North, 'Astronomy and mathematics', in J. I. Catto and T. A. R. Evans (eds), *The History of the University of Oxford*, II: *Late Medieval Oxford* (Oxford, 1992), p. 173.

9   J. M. Fletcher, 'The Faculty of Arts', in J. McConica (ed.), T*he History of the University of Oxford*, III: *The Collegiate University* (Oxford, 1986), p. 172.

10   Except where otherwise stated, biographical information in this instance and

elsewhere in this chapter is derived from relevant entries in *The Oxford Dictionary of National Biography* (60 vols, Oxford, 2004).

11   Sears Jayne, *Plato in Renaissance England* (Dordrecht, 1995), pp. 84–91.

12   D. R. Leader, *A History of the University of Cambridge*, I: *The University to 1546* (Cambridge, 1989), pp. 242, 249.

13   Ibid., pp. 289, 313.

14   Quoted in Kusukawa, *The Transformation of Natural Philosophy*, p. 36.

15   B. P. Copenhaver and C. B. Schmitt, *Renaissance Philosophy* (Oxford, 1992), p. 233.

16   For instance, the names 'Philotheus' and 'Theophilus' which Recorde lists as designating separate individuals in fact both refer to Theophilus Protospatharius, seventh-century AD Byzantine physician and compiler of a commentary on Hippocrates' *Aphorisms*, which was derived largely from Galen.

17   On Fine, see Alexander Marr (ed.), *The Worlds of Oronce Fine: Mathematics, Instruments and Print in Renaissance France* (Donington, 2009).

18   *Posterior Analytics* 74b27; on inception for the MA see Fletcher, in McConica, *The History of the University of Oxford*, p. 337. But cf. Recorde's own practice, notably in the *Pathway*, where conclusions and theorems in geometry are presented as 'Approued truthes', while the 'demonstrations and iust profes' are 'omitted, vntill a more conuient time' (*Pathway*, sig. [a.i]).

19   Pierre de La Ramée (Petrus Ramus), *Remonstrance de Pierre de La Ramée faite au Conseil privé, en la Chambre du Roi au Louvre, le 18 janvier 1567, touchant la profession royale en mathematique* (Paris, 1567), pp. 6–13:

> L'ordre des mathematiques n'est point comme d'une histoire, là ou vous pouuez entendre & declairer un passage à la fin, au meillieu, au commencement sans rien entendre au precedent: mais en la mathematique l'ordre y est non seulement profitable & utile, ains totallement necessaire: la premiere de ces disciplines c'est l'arithmetique, art de bien nombrer toute chose qui peut tomber en nature de nombre en adioutant, deduisant, multipliant, diuisant, tous nombres entiers & rompus en comparant leurs raisons & proportions; la seconde c'est la geometrie, art de bien mesurer, toute chose subiecte à mesure, comme longueur, largeur, hauteur, & generalement toutes grandeurs tant plaines que solides, soit au ciel, soit en terre ou en quelque autre subiet mesurable; ceste partie seconde ne se peut aucunement entendre ny praticquer sans la premiere; car mesurer c'est nombrer les interualles, c'est comparer les raisons & proportions des figures; l'astrologie qui s'ensuit, ne se peut pareillement ny conceuoir, ny demonstrer sans l'arithmetique & geometrie; car l'astrologie n'est autre chose qu'arithmetique à nombrer les degrez, minutes, & toutes autre parties es mouuemens des corps celestes, ce n'est autre chose que geometrie à mesurer les triangles, les cercles, les spheres & toutes figures y estant, & ainsi des autres disciplines mathematiques, voire bien daduantage les propositions d'arithmetique,

geometrie, astrologie sont basties de telle ordre, que qui ne cognoit la premiere, ne peut entendre la seconde, qui n'entend l'vne & l'autre ne peut entendre la troisiesme, bref si vn escolier a perdu vne seule leçon en mathematique, qu'il ne retourne plus à l'escole, car il n'entendra rien à ce qui s'ensuit . . . Quiconque commence les mathematiques contre l'ordre des mathematiques, en cela il declaire son insuffisance & son ignorance.

Cf. N. Bruyère, *Méthode et dialectique dans l'œuvre de La Ramée: Renaissance et âge classique* (Paris, 1984), pp. 360–1.

20 See Peter Mack, 'The dialogue in English education in the sixteenth century', in M.-T. Jones-Davies (ed.), *Le Dialogue au temps de la Renaissance* (Paris, 1984), pp. 189–212.

21 John Bale, *Index Britanniae scriptorum: John Bale's Index of British and Other Writers*, ed. R. L. Poole and M. Bateson, reissued with Introduction by C. Brett and J. P. Carley (Cambridge, 1990), p. 61; cf. T. W. Baldwin, *William Shakspere's Small Latine and Lesse Greeke* (Urbana Ill., 1944), vol. 2, p. 692.

22 Cf. R. T. Gunther, *Early Science in Oxford* (Oxford, 1923), vol. 1, Part 2, pp. 21–2, evidently deriving his 'catalogue of Recorde's library' from Bale without acknowledgement while omitting several titles. For further information see Bale, *Index Britanniae scriptorum*, p. xxxi.

23 On Prise, see N. R. Ker, 'Sir John Prise', *The Library*, 5th ser., 10/1 (March 1955), 1–24.

24 Sir Thomas Elyot, *The Boke Named the Gouernour* (London, 1539), I.xv (sigs h.vᵛ–h.vj).

25 John Dee, *The Mathematicall Praeface to the Elements of Geometrie of Euclid of Megara* (London, 1570; facsimile edn, New York, 1975), sig.*j.

26 I owe this point to the kindness of Dr John Watts in allowing me to cite from his unpublished paper 'Common Weal to Commonwealth'; cf. his 'Public or plebs: the changing meaning of the "Commons", 1381–1549', in Huw Pryce and John Watts (eds), *Power and Identity in the Middle Ages: Essays in Memory of Rees Davies* (Oxford, 2007), pp. 242–60.

27 A. F. Pollard, *England under Protector Somerset: An Essay* (London, 1900), pp. 200ff.

28 Whitney R. D. Jones, *The Tudor Commonwealth, 1529–1559* (London, 1970), pp. 24, 32 for use of the term 'party' in this connection.

29 G. R. Elton, *Studies in Tudor and Stuart Government and Politics*, vol. III: *Papers and Reviews 1973–1981* (Cambridge, 1983), pp. 234–8.

30 Elton's dismissive term: *Studies*, vol. III, p. 39.

31 Thomas Starkey, *A Dialogue Between Pole and Lupset*, ed. T. F. Mayer, Camden, Fourth Series, 37 (London, 1989), p. 38.

32 John Hooper, *A Declaration of the Ten Holy Comaundementes of Allmygthye God* (Zurich, 1548), pp. CXXII–CXXIII.

33 King Edward VI, *The Chronicle and Political Papers of King Edward VI*, ed. W. K. Jordan (London, 1966), p. 165.

34   Leonard Digges, *A Geometrical Practise Named Pantometria* (London, 1571), sig. * iv^v.

35   *The Grounde of Artes* (1582), sig. A.ij^v.

36   Richard Benese, *This Boke Sheweth the Maner of Measurynge of all Maner of Lande* (London, 1537), sigs +ii, [+iiii].

37   Richard Eden, *The Arte of Nauigation, Conteynyng a Compendious Description of the Sphere* (London, 1561), sig. Cii; translated from Martin Cortes, *Breue compendio de la sphere y de la arte de nauegar* (1551).

38   John Dee, *General and Rare Memorials Pertaynyng to the Perfect Arte of Navigation* (London, 1577), pag. 10; cf. W. H. Sherman, *John Dee: The Politics of Reading and Writing in the English Renaissance* (Amherst, 1995), p. 150.

39   Dee, *General and Rare Memorials*, pags 3, 8.

40   B. W. Henry, 'John Dee, Humphrey Llwyd, and the name "British Empire"', *Huntington Library Quarterly*, 35 (1971–2), 189–90.

41   See above, n. 1.

42   W. Ullmann, 'On the influence of Geoffrey of Monmouth in English history', in C. Bauer, L. Boehm and M. Müller (eds), *Speculum historiale: Geschichte im Spiegel von Geschichtsschreibung und Geschichtsdeutung* (Munich, 1965), pp. 257–63.

43   Quoted in G. Nicholson, 'The Act of Appeals and the English Reformation', in C. Cross, D. Loades and J. Scarisbrick (eds), *Law and Government under the Tudors* (Cambridge, 1988), p. 24.

44   *The Historia Regum Britanniae of Geoffrey of Monmouth*, ed. Acton Griscom and Robert Ellis Jones (London, 1929), pp. 446, 455.

45   Sherman, *John Dee*, p. 188, citing Dee's *Famous & Riche Discoveries*.

46   Quoted in Gwyn A. Williams, *Madoc: The Making of a Myth* (London, 1979), p. 39.

47   Robert Fabyan, *The New Chronicles of England and France in Two Parts*, ed. Henry Ellis (London, 1811), pp. 19–20. Fabyan himself had died in 1513.

48   On links between the 1565 play *Gorboduc* by Thomas Norton and Thomas Sackville and Shakespeare's tragedy about yet another legendary British king and supposed ancestor of Gwrvyw, see for instance Barbara Heliodora Carneiro de Mendoça, 'The influence of Gorboduc on King Lear', *Shakespeare Survey*, 13 (1960), 41–8. In *Lear* Merlin and prophecies are cited somewhat anachronistically in a speech of the Fool's (*Lear* III.ii.95).

49   *Whetstone*, sig. a.iii^v; *Castle*, p. 213 [212]; cf. E. G. R. Taylor, *Tudor Geography 1485–1583* (London, 1930), p. 24.

50   The reference must be to Richard Chancellor's 1555 voyage to Russia – the Muscovy Company's first venture following the grant of its charter in February of that year (T. S. Willan, *The Early History of the Russia Company, 1553–1603* (Manchester, 1956), pp. 10, 7).

51   L. D. Patterson, 'Recorde's cosmography, 1556', *Isis*, 42/3 (October 1951), pp. 209–10.

52   Sherman, *John Dee*, p. 175.

53   K. R. Andrews, *Trade, Plunder and Settlement: Maritime Enterprise and the Genesis of the British Empire, 1480–1630* (Cambridge, 1984), p. 167, n. 2; Taylor, *Tudor Geography*, p. 94.

54   Taylor, *Tudor Geography*, p. 93; Antonia McLean, *Humanism and the Rise of Science in Tudor England* (London, 1972), p. 133.

55   Amir R. Alexander, *Geometrical Landscapes: The Voyages of Discovery and the Transformation of Mathematical Practice* (Stanford, 2002), pp. 1–2.

56   Boyer, *A History of Mathematics*, p. 306.

57   See n. 18.

NINE

Data, computation and the Tudor knowledge economy

JOHN V. TUCKER

But thou hast arranged all things by measure and number and weight.

Wisdom (Apocrypha) 11:20

Introduction

FROM HIS BIRTH IN TENBY *c.*1510 to his death in Southwark in the summer of 1558, Robert Recorde lived through 'interesting times':[1] the turmoil of the reigns of Henry VIII, Edward VI and Mary. Elizabeth was crowned some six months after Recorde's death and her reign saw important events and changes in which we can trace Recorde's influence.

In this chapter I consider the historical and intellectual context of Robert Recorde's work in terms of the new role that mathematical and computational ideas and methods were beginning to play in the society and economy of Europe. The sixteenth century saw trade in new international markets and products; increasing availability of information through printing; an increase in travel, mobility and the development of international networks; and an intensification of technical and expert knowledge in practical activities. The Tudor period saw a European society and economy becoming heavily dependent on measurement and calculation in its organization and activities, stimulating the rise of practical mathematics. The collection, generation and use of data, broadly conceived, changed the way in which knowledge and money were employed and disseminated in Europe.

Over the period of Recorde's lifetime, the British Isles lagged behind continental Europe in the development of commerce, in the establishment of educational institutions and, more particularly, in the exploitation and extension of mathematical ideas. Italy set the standard, with its banks and companies, its universities and schools, and the blossoming of an extensive mathematical literature. Developments in Italy were spreading to Germany, the Low Countries and France. Recorde's series of books in the vernacular, beginning with *The Ground of Artes*, aimed at teaching contemporary mathematics in order to improve practical professions. It succeeded in that aim and was, therefore, influential in 'modernizing' the Tudor period. Indeed, one can argue that Recorde's series is an important intellectual milestone in the history of modern Britain.

Recorde is best remembered for his mathematical works. However, his contribution does not stand tall in writings about the history of mathematics, which are fundamentally technical histories of pure mathematics. He founded modern mathematics in Britain, but his works are expository; they are orientated towards practical activities that are removed from pure mathematics; and they are based on mature scholarship rather than original discoveries. To appreciate and celebrate Recorde one needs to be interested in more than the technicalities of mathematics: one needs to be interested in the dialogue between mathematics and the world's work.

However, the view from the perspective of computer science, my own field, is different. Computer science is a mathematical science that is much closer to the world's work. Computing is largely about collecting, creating and processing data. It is universal and ubiquitous because it is *intimate* with the world's work, which is based upon gigantic quantities of data. Thus, in seeking the origins of computing, the essential ideas are *quantification* and *data*. To understand the history of computing, we may follow the data, which leads us to practical mathematics and to writers such as Recorde. Data connect computing directly with knowledge, expertise and professional practice.[2] In the Tudor period the growing awareness of this connection led to new conceptions of science and the widespread education of people in the ways of mathematics.

In this chapter I offer some reasons for taking an interest in Recorde that arise from the history of computer science. In Section 1, I take a view on scientists engaging with the history of their subject before reflecting on the contemporary themes of data, computation, knowledge and expertise that suggest my approach. In Section 2, I consider the commercial revolution in Europe linked with social change and the new and abstract ways of doing business using international banks and such tools as letters of credit and bills of exchange. In Section 3, I comment on Recorde's own involvement in contemporary affairs. Finally, in Section 4, I draw attention to the relevance of the achievements of Recorde and his contemporaries in debates about our understanding of the development of science and the modern world.

1. Views of Recorde from the present

We write and rewrite history to keep the past alive. Each generation has different knowledge, experience, perspectives, curiosities and questions. Historians have problems, tastes and prejudices shaped by contemporary agendas. For the past to speak to the present is incredibly difficult and requires the patient work of generations of scholars. Their scholarship creates historiographies and sustains history as a discipline. The question endures: Here we are, what is our history?

The dialogue between the present and the past is evident in the process of interpreting the history of science and technology. This process commonly involves two sets of people and their characteristic expertise: historians who know about periods, places, events, people, networks and historical methods; and scientists and engineers who know about the theory and practice of their subject and are interested in history. This dichotomy is inevitable in the history of technical subjects.[3]

The history of science and technology is now sufficiently mature as an academic discipline to have nurtured scholars that combine both historical and scientific competences. Historians seek the past as it was and give old technical ideas and perspectives the care and attention they need. They maintain an intellectual development dependent on the traditions of writing history and independent of contemporary science and technology. They can be weak on new technical subjects and perspectives. Scientists and engineers seek origins of contemporary topics, sometimes with an intention of using history to progress their science and its reception. They can be weak on old technical subjects and perspectives. The history of science and technology is intriguing intellectually because it seems to be practically dependent on philosophical questions. The pressure both to maintain contemporary relevance and to conform to a particular philosophical conception of what constitutes science can lead to selectivity in historical studies. One thinks, for example, of the deliberate neglect of the study of alchemy over many decades.[4] Historians are better able to control such philosophical influences.

Cultures and identities play a role, too. Much of the history of science and technology is international, focused on the pursuit of ideas, methods and achievements that are independent of people and places. Literally *hundreds* of scientists have made important and significant contributions to science. Sometimes a hard line is taken and there is no shortage of descriptions of outstanding scientists being of the 'first' or 'second' rank. Such an attitude is often taken in the history of mathematics, which is dominated by the history of pure mathematical concepts and theorems.[5] In the British Isles, Recorde is the most distinguished mathematician of his age but a hard-line historian of mathematics may wonder: What new mathematical ideas, methods and achievements are there to expound, develop and celebrate?

We rightly expect views of Recorde from historians of mathematics, of education, of scholarship, of commerce and of the Tudor period in general. Until recently, these views were somewhat limited in their scope.[6] I am a computer scientist interested in the history of my subject and offer a view of Recorde framed by the history of computer science, which is an emerging field within the history of science and technology. Computer scientists have a legitimate interest in Recorde's times and culture because of the role of quantification, data, computation and technical education, both in computing science and in knowledge economies.

1.1   Data and computation

Computer science is about data and computation. It studies

(i)   representation and storage of data
(ii)   algorithms for transforming data and creating new data
(iii) programming methods for constructing software to represent algorithms, and
(iv)   methods for designing and operating machines and networks of machines to implement software.

Computer science is a mathematical science but it is not mathematics. Computing involves mathematical models and programming; the models draw on all parts of mathematics and the programming is based on arithmetic, algebra and logic.

The history of computer science includes the history of these four components, (i)–(iv). In terms of components (i) and (ii), that history spans a period of over 2,000 years as exemplified by the algorithmic methods developed by Greek mathematicians, notably Euclid's algorithm to calculate the greatest common divisor of two numbers and Eratosthenes' algorithm in the form of a sieve to locate prime numbers. However, the full manifestation of computer science brings to mind images of modern computers and the profound influence they have had on almost every part of life through their seemingly limitless practical applications. The history of computer science is also about civil and military applications; the creation of vast new businesses, organizations and government; changes in education and skills, and in social and cultural behaviour. These are the events of a single lifetime: if you were born in or before 1936, the year of Alan Turing's discoveries, you will have lived through it all.[7]

Data are created and used to represent, reason and manage the world. Data are a commodity. Today's world is awash with data: scientific data (e.g. in astronomy and meteorology); medical data (e.g. 3D NMR scans, patient records); engineering data (e.g. plans, measurements); security and surveillance data (e.g.

videos, communication logs); persuasive data (e.g. as collected by traffic cameras, public CCTVs); media data (e.g. music, television programmes, films, news); archives (e.g. documents, papers, books, photographs, videos); business data (e.g. accounts, inventories, customer data, contracts, credit ratings); government data (e.g. censuses, taxes, licences, identities); personal data (e.g. emails, blogs, albums). Finally, there is money, to which so very much of the world can be reduced; truly, money is the ultimate data type for quantification.

The history of computing is necessarily, and obviously, socio-technical, in that it cannot be a history of technicalities independent of the world. In search of the origins of computing we encounter the world in familiar ways; technical problems to do with data, representations and algorithms are close to the world's work for which quantification, data, information and knowledge are the conceptual *sine qua non*. Since 'data' is the primary concept of computer science, the question arises: What are the origins of our use of data and our dependency on data? In search of answers to these big questions we encounter Recorde and his world.

Those aspects of our world that have been numbered, weighed, measured or costed in some way have grown more numerous over the millennia. The rise of practical mathematics, its wide application, its fathering of specialist fields such as accounting, and its evolution into algebra mark the beginnings of new uses of mathematics. They also mark a rise in the pursuit of quantification, in which data are facts created and used to represent, understand and manage the world. Practical mathematics in the sixteenth century is an important part of the history of computer science. Recorde is at the heart of this transformation in the British Isles.

1.2 Knowledge and the economy

Since about the year 2000, the phrase 'knowledge economy' – if not always the related concept – has been used extensively in public discussion and policy development at all levels of western societies.[8] Originally formulated by Peter Drucker (1909–2005), the idea has developed and changed along with our understanding of policies.[9] For example, an anonymous author at the UK Economic and Social Research Council has offered this gloss to aspiring researchers:

> In today's global, information-driven society, economic success is increasingly based upon the effective utilisation of intangible assets such as knowledge, skills and innovative potential as the key resources for competitive advantage. The term knowledge economy is used to describe this emerging economic structure and represents the marked departure in the economics of the information age from those of the twentieth-century industrial era.[10]

Rhetorically, the special characteristics of knowledge economies include:

- globalization of markets and products due to national and international deregulation
- increasing availability of information and communications technologies
- increasing networking and connectivity over the internet, and
- intensification of knowledge in economic activities, by information technology growth and high-tech products and services.

Computer science is the science that gives us our understanding of data and information. I use the term 'knowledge economy' loosely, as a fashionable nickname to emphasize the analysis of knowledge and the economy. Although it belongs to a contemporary scene, aspects of the idea resonate with Tudor times and with the world of Robert Recorde.

2. Computation and data and the early Tudor period

As we have noted above, at the beginning of the sixteenth century Britain lagged behind the rest of Europe, notably Italy, in terms of its commerce, its educational infrastructure and, in particular, its development of mathematical ideas. Renaissance Italy was the result of a commercial revolution starting in the thirteenth century,[11] which led to social change and business innovations, coupled with a flowering of art and architecture, of curiosity and investigation. Recalling the characteristics of knowledge economies set out above, we may describe the transformations in sixteenth-century Europe as involving:

- trade in new international markets and products
- increasing availability of information
- increasing travel, mobility and international networks, and
- intensification of knowledge in economic activities.

Italy had innovated along these lines, developing a range of new techniques and institutions, including: international banks; letters of credit and bills of exchange; underwriting and insurance; continuous accounting with double-entry bookkeeping; loan interest and profit dividends; schools for calculation; universities; and, printing presses with wide-ranging catalogues. Italian business was becoming increasingly abstract in its ability both to collect and generate data and to compute with those data. What data were involved and what were the computations? Who made these computations and how did they learn their methods?

2.1 Italian abbacus manuscripts and books

Commercial computation was essential practical knowledge in the developing Italian economies. It was taught in schools to 8–10-year-old children by *maestri d'abbaco*. Computation was the subject of a vernacular manuscript and book tradition based on *libri d'abbaco* – abbacus texts. The tradition descends from Leonardo of Pisa's *Liber abbaci* (1202), with its origins in the medieval Arab world.[12]

The mathematical tradition at the beginning of the sixteenth century was captured by Luca Pacioli (1445–1517) in his great work *Summa de arithmetica, geometria, proportioni et proportionalità*, published in Venice in 1494, with a second edition in 1523. Written in Italian, it provided comprehensive coverage over 600 pages of: arithmetic; elements of algebra; tables of moneys and weights; double-entry bookkeeping; and Euclidean geometry. Later Italian mathematicians, including Girolamo Cardano, Niccolò Tartaglia, Lodovico Ferrari and Rafael Bombelli – all familiar figures in the history of algebra – felt the influence of this work. Pacioli also had an enduring influence on accounting.

But it is the nature of the hundreds of published *libri d'abbaco* that we should consider. The abbacus texts have been studied in depth by van Egmond who has revealed an important mathematical tradition.[13] Their distinctive features are that they use:

- Hindu–Arabic number systems exclusively
- modern methods of calculation
- large collections of sample problems
- wide varieties of problems
- practical situations as examplars
- meticulous step-by-step explanations, and
- algebraic method.

(Somewhat confusingly, abbacus texts have nothing to do with the abacus.) Van Egmond has observed a representative structure for the tradition:

- Preliminary material: Hindu–Arabic numbers; addition, subtraction, multiplication and division, using whole numbers and fractions; moneys, weights and measures.
- Business problems: pricing; money exchange; weights and measures; barter; partnership; interest; discount; loan repayment; alligation.
- Recreational material: number problems, number problems in disguise; series and progressions.

- Geometry: abstract geometric figures; measurement problems based on real objects.
- Methodology: Golden rule; solution of algebraic equations.
- Miscellany: calendars; tariffs.

The abbacus texts and schools constitute an important achievement. Firstly, they modified and improved the Arabic number system and calculation methods to meet Western needs, providing a basis for our modern methods. While it is the case that Hindu−Arabic numbers were to be found in Italy from the time of Leonardo of Pisa, the methods for performing arithmetic operations with these numbers changed markedly. Arabic methods for multiplication and division used the special features of sand trays; they were not fully symbolic and could not be written down with pen and paper. Indeed, detailed examination of number systems leads to a case to rename the Hindu−Arabic numbers as simply *western* numbers.[14] By the early sixteenth century the *maestri d'abbaco* had developed our modern fully symbolic rules.

Secondly, the abbacus texts and schools transformed the conduct of business in Europe by introducing and embedding mathematics into commerce and society. Trade between Italy and northern Europe brought about the development of abbacus traditions in Germany and the Low Countries.[15]

Thirdly, they provided a huge technical and cultural impetus for the development of algebra; the practical problems that filled the texts cried out for general algebraic methods. The abbacus texts in turn stimulated the composition and publication of new mathematical texts across Europe. The abstractness and generality of algebra is a fundamental attribute of mathematics and its use. Only occasionally do mathematicians remember that, in Bochner's phrase, '. . . the challenges, to which the rise of algebra was the response, were, predominantly, the very unlofty and utilitarian demands of counting houses of bankers and merchants . . .'[16]

2.2 Developments in the British Isles

As data and computation advanced in mainland Europe, what was the situation in the British Isles?

There is the extraordinary picture of Hans Holbein's *The Ambassadors*. Painted in London and dated 1533, it depicts two young Frenchmen, Jean de Dinteville, aged twenty-nine, and his friend Georges de Selve, the bishop of Lavaur, aged twenty-five. If Recorde was born in 1510, he would have been twenty-three at the time. The image is a rich representation of a mercantile Tudor world. Beneath a globe, we can see a book on arithmetic, partly open on a page of long divisions. Compiled for use by merchants, written in German by Peter Apian (1495–1552) and published in 1527, the book is a practical guide on calculation. *The Ambassadors* is a highly complex image whose iconography is well suited to our subject.[17]

What other arithmetics may have been accessible to Recorde during his formative years? In 1522 Cuthbert Tunstall (1474–1559) published *De arte supputendi*, an arithmetic in Latin. Tunstall had mathematical training and had resided in Padua during 1499–1505, where it is assumed that he became familiar with Italian arithmetic. In the same year, Tunstall became bishop of London and began outlawing William Tyndale's English translation of the New Testament. The attachment to Latin scholarship is noteworthy.

Commercial arithmetics in English were slow to appear: an English translation of Pacioli by Hugh Oldcastle (d. 1543), published in London in 1543, is now lost. The next native English book on accounting, based on Oldcastle and hence Pacioli, appeared forty-five years later: John Mellis's *A Briefe Instruction and Maner How to Keepe Bookes of Accompts after the Order of Debitor and Creditor* (London, 1588). Mellis (*fl.* 1564–88) was a schoolmaster who, from 1582 onwards, produced, revised and extended editions of Recorde's *Ground*.

Among English merchants, Thomas Gresham (*c*.1518–79) was eminent and fabulously wealthy. During his period, the Crown regularly needed loans and, since there were no bankers, merchants acted as financiers arranging loans from abroad. Gresham served as financier for Edward VI, Mary and Elizabeth. The nature of his business activities required him to commute for months each year to Antwerp, where he raised loans for the Crown at the Bourse. Gresham's accounts for 1546–52 were the first in Britain to use double-entry methods, although the methods were not fully exploited.[18]

In order to trade efficiently, merchants needed a thorough understanding of money and value, complicated by the existence of a multiplicity of systems of coinage and currency, developed independently by individual cities and centres of commerce. There were also many competing systems of weights and measures, made more complex by being dependent on products: a gallon was not a standard volume of fluid but varied according to the fluid that was being gauged. International trade compounded these commercial difficulties. The growth in international trade and the huge ambitions of naval and merchant adventurers generated a need for navigation instruments and maps.[19] The dissolution of the monasteries and the disposal of vast lands and the creation of large building projects generated a need for surveying.[20]

During the period of Recorde's youth there was therefore plenty to measure and calculate. However there was also a dearth of sources in English readily providing even elementary knowledge. The point is clearly made in Table 1, which compares the number of printed texts during this period in Italy and Britain, based on Van Egmond's catalogue of in excess of 300 printed Italian Arithmetics.

Table 1. Printed abbacus texts from 1470 to 1560

Decade	Number of printed abbacus texts	Italian highlights	British highlights
1470	2	Anon, *Treviso arithmetic*, 1478	
1480	2	Borghi, *Arithmetica*, 1484	
1490	4	Pacioli, *Summa*, 1494	
1500	2	Pacioli, *Divina proportione*, 1509 Pacioli, *Euclidis Margarensis*, 1509	
1510	9		
1520	20		
1530	12	Tartaglia, *Nova scientia*, 1537 Cardano, *Practica arithmetice*, 1539	Anon, *An Introduction for to Lerne to Rekyn with the Pen and with Counters*, 1536/7
1540	18	Cardano, *Ars magna*, 1545 Tartaglia, *Euclide Megarense*, 1543 Tartaglia, *Questi et inventioni*, 1546	Recorde, *The Ground of Artes*, 1543
1550	15		Recorde, *The Pathway to Knowledg*, 1551 Digges, *A Booke Named Tectonicon*, 1556 Recorde, *The Castle of Knowledge*, 1556 Recorde, *The Whetstone of Witte*, 1557 Cuningham, *The Cosmographical Glasse*, 1559
1560	19		

2.3   Recorde's books

Commentaries on Recorde's individual extant works on mathematics – *The Ground of Artes, The Pathway to Knowledg, The Castle of Knowledge* and *The Whetstone of Witte* – are to be found elsewhere in this volume. Let me summarize some of their special features that are relevant to our argument.

First and foremost, Recorde's mathematical books form a *programme of instruction* that take the reader on an intellectual journey. The works are introductory, in English, and are designed to teach some advanced ideas to a wide audience. They present a coherent body of mathematical thought in which numbers are abstract and explanations are important. Recorde had a particular vision of the nature of knowledge – that it is:

- broadly and practically conceived and includes commerce, land surveying and navigation
- precise and quantitative, and
- firmly based upon sound reasoning that must be open to demonstration and even dispute.

Furthermore, Recorde had a vision that arithmetic and geometry are a foundation for knowledge. Thus, Recorde introduces the Tudor reader not just to mathematical methods but to a new mathematical frame of mind and a new conception of, and personal relationship with, knowledge.

These four books are the foundation of a large programme on mathematics and its applications in the world's work. They contain observations on the nature of knowledge and on the fundamental role of measurement, calculation and reasoning. Educating the reader in essential mathematical ideas and methods, they are a preparation for practical applications. The title page of the *Pathway* emphasizes the book's objective of providing the reader with sufficient knowledge to use geometrical and astronomical instruments, and the book's Preface extols the virtues of geometry in a range of practical applications, so that, for example, 'if he kepe not the rules of Geometrie, he can not measure any ground truely' (*Pathway*, sig. [s.iv]^v). Verses that appear in the *Castle* (sig. [a.viii]) suggest that the third of Recorde's planned mathematical texts was *The Gate of Knowledge*, focusing on practical geometry and measurement by the quadrant.[21] Sadly, no copy of the *Gate* has been found. In the 1552 edition of the *Ground*, a future work is mentioned on coinage.

In the years following Recorde's death, interest in mathematics grew considerably. The Elizabethan period saw significant achievements involving mathematical and computational methods and their application. The mathematical tradition includes landmarks such as John Dee's *Praeface*, attached to Billingsley's 1570 translation of Euclid (see chapter 4); and Leonard Digges's treatise in 1571 on measurement.[22] The period also witnessed a significant growth of mathematical and scientific activity of a practical nature.[23]

The technical ambitions of the navy and merchant marine provide numerous illustrations. Recorde's *Pathway* had shipbuilding and the use of navigational instruments in mind. The *Whetstone* was dedicated to the 'venturers into Moscouia', the Muscovy Company, subsequently served by John Dee, being an association of merchant adventurers (see chapter 6). Indeed, through Dee's mathematical interests in navigation some of the knowledge of the pioneering European navigators and map-makers were communicated to the leading British sailors.

There developed a significant literature on navigation. Texts such as William Borough's *A Discovrs of the Variation of the Cumpas* (London, 1581) are addressed to experienced mariners but contain complex arithmetic and geometry. Borough claims that a sufficient number of English mathematical texts were available to enable practitioners to master the prerequisites.[24]

In the seventeenth century, the market for arithmetic and geometry expanded considerably. However, the excellent standards of exposition and explanation achieved by Recorde began to decline. There was a corresponding decline in scholarship and well-founded mathematical argument in favour of shortened, more superficial manuals. For example, the emergence of ready reckoners testifies both to the importance and popularity of computation and the expectation that arithmetical understanding is an unnecessary trouble, distraction or luxury. As explanations were jettisoned and texts were customized to specific tasks or professions, interest in the universal nature of the methods and their applicability disappeared, certainly from the attention of mathematicians and the academy. Recorde's mathematics is an early example of a familiar phenomenon: a rigorous technical subject, as it becomes widely employed, is destined to be dumbed down. More is less.

3. Money

Recorde was a highly educated and learned scholar with a wide range of intellectual interests. He was a physician, mathematician and administrator. Known at court and active in the nation's affairs, he was a scholar heavily involved in the world's work. In particular, he was intimate with the world's favourite form of data, namely moneys.

Recorde's involvement with the Royal Mint covered a short but important period of his life. It began in 1548, some five years after he wrote *The Ground of Artes*. Coinage and currency are both practical and theoretical subjects par excellence at any time, but involvement with them during this period presented particular difficulties. Through his work with the mint, Recorde was to be caught up in the political events of 1549.

The currency was debased toward the close of the reign of Henry VIII. The king was Treasurer, and coinage production increased with the establishment of new royal mints at Bristol (1546) and, under Edward VI, at Durham House (1548). The new Durham House mint in the Strand was commissioned in December 1548, with John Bowes as its under-treasurer and Robert Recorde as comptroller. Its output was small and was thought to be a benefit for Protector Somerset. The Bristol mint had closed in 1472 and was reopened in 1546, with William Sharrington (c.1495–1558) appointed as its under-treasurer. Sharrington was at Henry's court. Knighted on the day of Edward VI's coronation, in February 1547, Sharrington was a well-connected figure in the royal mint.[25]

Sharrington's tenure was lucrative and eventful. He had bought Lacock Abbey at its dissolution in 1539 for £763. He made a fortune at the mint by reducing coin sizes and pocketing the difference. Sharrington entered into a financial association with Sir Thomas Seymour, the difficult younger brother of Protector Somerset. When Seymour's activities became serious problems for the Protector, he was accused of treason. In consequence, Sharrington and the Bristol mint were investigated by Sir Thomas Chamberlayne.

In January 1549 Sharrington was arrested. Sir Thomas Chamberlayne promptly succeeded him as under-treasurer with Robert Recorde as comptroller. In March, Seymour was executed. In May, Sir William Herbert tasted the people's opposition to the land enclosure policy when his park at Wilton was attacked. By June, Chamberlayne was ambassador in Denmark leaving Recorde in charge at Bristol. In the summer of 1549, Herbert and Russell were part of the campaign to put down the Prayer Book Rebellion in Devon and Cornwall. By August, the rebellion was over.

In October, Herbert demanded money from the Bristol mint for his campaign expenses. Recorde refused, requiring instruction from the king. Herbert caused Recorde to be recalled to London and the Bristol mint to be closed. In the same month the disturbing events of the year – culminating in public rebellions – led to Somerset's dismissal as Protector and his imprisonment.

Sharrington was pardoned in November and allowed to purchase his freedom for £8,000, subsequently regaining his property. Sir William Herbert and the Earl of Warwick were made chief commissioners of the king's mints in February 1550. However, Recorde continued to work with the royal mint in Ireland during 1551–3.

The mathematical basis of coinage in the period from Leonardo of Pisa to Robert Recorde has been analysed by Jack Williams, who has established connections between currency and the abbacus texts and later arithmetics. In particular, he traces the influence of Recorde's work with coinage on his 1552 edition of the *Ground*, to which Recorde added fractions and methods of alligation and mixing, illustrated with related examples.[26]

Recorde accused Herbert of malfeasance in a letter of June 1556.[27] Herbert sued Recorde for libel, claiming damages of £12,000. At the trial, in February 1557, Herbert was awarded £1,000 which brought Recorde to the King's Bench Prison, near the Borough Road, Southwark, where he died.

Sir William Herbert (1506–69) is an interesting figure.[28] His political survival at the highest levels of government, from the courts of Henry VIII to Elizabeth, is remarkable. William married Ann Parr in 1534. He became the king's brother-in-law on Henry VIII's last marriage, to Ann's sister Catherine in 1543. Family relationships became more involved when the dowager Queen Katherine married Sir Thomas Seymour. Herbert rose, survived and prospered in money, power, grandeur and rank, and laid the foundations of a great family. Herbert was a violent man and a hardened soldier. His support was needed for changes in power, including the removal of Somerset and the accession of Mary. Precisely why Recorde quarrelled with Herbert is difficult to fathom.

4. Debates on the origins of science and modernity

Recorde founded mathematics for practitioners in the British Isles. His works were rigorous and based on sound pedagogical principles; for many years they were the primary source of knowledge in the field of mathematical methods and calculation, influencing the way in which mathematics and its application were perceived and conceptualized. The intellectual foundation that he provided made it possible for mathematics and mathematical methods to be at the core of the development of the Elizabethan economy into the seventeenth century. The nature of Recorde's work secures him a place in some remarkable debates regarding the history of intellectual development. First, let me try to indicate how Recorde features in debates on the origins of modern science.

4.1 Origins of modern science

Scientists, historians and intellectuals of many kinds are interested in the origins of science. Science is a way of knowing and acting on the world, one that has emerged relatively recently and exclusively in Europe. Thinking about science historically involves formulating a philosophy of knowledge, as one has to establish some 'essential characteristics' of science with which to navigate through the intellectual past. Searching for the origins of science involves identifying its characteristics. As science changes, and interest in its essential characteristics changes, so do our questions concerning its nature and our understanding of its essence.

For most of the twentieth century, the essential characteristics of science – encompassing the nature of the scientific method and the purpose of science within society – that shaped scholarship on the origins of modern science included:

(i)   observation as a means of collecting data
(ii)   systematic experimentation
(iii)   usefulness of knowledge
(iv)   measurement
(v)   calculation
(vi)   mathematical theories.

These last three characteristics, (iv)–(vi), together constitute what is commonly called the 'mathematization of nature', which involves creating and measuring quantities and qualities, calculating new data, and modelling and simulation to explain and predict behaviours. To all these may be added other characteristics such as:

(vii)   new objects of scientific study, such as commercial applications and psychic powers
(viii) independence from religion
(ix)   new social infrastructures.

This last characteristic, (ix), includes a number of subcategories: collaboration and networks; societies and meetings; authority and advice; laboratories; secrecy and intellectual property; technical education; and patronage.

By selecting and combining some of these characteristics and analysing their nature and historical origins, a fascinating literature has grown that emphasizes some areas and neglects others. However, the literature on the nature of science is rather complicated and one soon finds a need for doses of historiography and synthesis to appreciate the arguments.

Recorde's work assumes observation (i) and measurement (iv), and expounds calculation (v) for useful purposes (iii); and it introduces methods to a wide audience (ix). In these characteristics Recorde is not alone, for he is one of a number of early authors of superior mathematical works such as Pacioli, Cardano, Rudolff and Stevin. Recorde's work should be thought of as belonging to a wider European movement that made important advances in Italy, Germany and the Low Countries.

However, Recorde initiated practical mathematics and computation that had a sustained impact into the seventeenth century on a country that was becoming a world power and in which several great landmarks of science were being created, including such native highlights as the work of Francis Bacon (characteristics (ii) and (iii)) and of Isaac Newton (characteristic (vi)) and the founding of the Royal Society in 1660.

The importance of Bacon in the historiography of science is well established. The influences on him are less well known. Dee's *Mathematicall Praeface* anticipates aspects of Baconian theory and systematizes ideas to be found in Recorde's books. For example, Recorde's idea of the divine importance of number and measure is restated in Dee's *Praeface*, but it is not new, as the epigraph to this chapter shows. Analysis of the influence of Recorde on Dee and Bacon could form the basis of an interesting research problem. However, Bacon's philosophy was not mathematical, and therefore philosophical connections may prove to be subtle and historical evidence of such connections may be elusive.[29] Recorde's role in the development of science in the British Isles and elsewhere is not dependent on such simple connections.

The search for the origins and development of modern science is necessarily linked with Western Europe's profound impact on the world. In 1949 Herbert Butterfield first argued that the origins of modern science are to be found in the events of the seventeenth century, constituting a 'scientific revolution'.[30] The appeal of this notion and its influence on historical writing in the intervening years are considerable. But the notion of a 'scientific revolution' can distort both the nature of historical writing and our thinking about the essence of science. It is particularly misleading when used casually in popular discourse: for example, if modern science is thought to be a *creation* of the times of Galileo and Newton then, clearly, Recorde would be perceived to be remote and of little relevance. The very idea of 'scientific revolutions' led some historians of science into deep historiographical controversies, but more recent reappraisals of the notion have provided a restorative balance.[31]

The origins of modern science are the subject of Edgar Zilsel's classic paper 'The sociological roots of science' (1942), which invites an exploration of the social conditions of the period 1300–1600.[32] Among the ideas aired by Zilsel about the origins of modern science are:

- The principles of causal explanation and methodological experimentation derive from the working practices of craftsmen, artisans, surgeons, instrument makers, surveyors, navigators, engineers and architects; this idea is often referred to as 'Zilsel's Thesis'.
- The mathematical encapsulation of nature in the seventeenth century depends heavily on the commercially inspired mathematical work of Pacioli, Recorde, Digges, Tartaglia and others.
- The social context is the development of capitalism.

Some precursors of these ideas belong to Marxist histories of the 1930s.[33] In our context, we note that Zilsel's arguments make direct links to the work of Recorde and his contemporaries.

The origins of modern science pose a specific intellectual problem that can be simplified by choosing to think about its particular characteristics, such as those listed above, (i)–(ix). As noted, many of these characteristics may be found in Recorde's works. As a further example, one particular quotation from *The Castle of Knowledge*, (p. 127 [129]) now carved in slate,[34] reflects changes in attitude to the nature of knowledge and authority (cf. characteristic (ix)):

> . . . yet muste you and all men take heed, that . . . in al mennes workes, you be not abused by their autoritye, but euermore attend to their reasons, and examine them well, euer regarding more what is saide, and how it is proued, then who saieth it: for autoritie often times deceaueth many menne . . .

The scope, standard, purpose and public nature of Recorde's mathematical works exemplify several of the most important features of the emerging forms of scientific knowledge.

4.2 Origins of modernity

By 'modernity' we mean the essential characteristics of the modern world, itself a European creation. Crudely, modernity characterizes those forms of social life that emerged in Europe and have been exported across the world since the fifteenth century. Science is one such characteristic; industrial capitalism is another. Modernity is ultimately a sociological notion emanating from theories built from abstract concepts and designed to explore social phenomena of the modern world.[35] However, such sociological notions can be highly problematic when confronted by history. For example, historians have long employed relatively lightweight notions of 'modernization' in their narratives, knowingly and unknowingly, and not without difficulties.[36]

The pursuit of questions about the origins of things modern has led historians of ideas to enter the scholarly world of Recorde and his contemporaries. One particular feature of modernity is that of quantification, encompassing the systematic collection of measurements, records and other data and their analysis by calculation and computation.

Here is an example of a theory based on quantification that illustrates the point. Alfred W. Crosby considered the historical problem of how to understand the success of European imperialism and the dominance of European based societies. In his *The Measure of Reality*[37] he formulated the thesis that this success can be attributed to the development by Europeans of the capacity and the mentality both to organize large collections of people and capital, and to exploit physical reality in order to gain knowledge and power. The decisively important factors that determine capacity are the administrative, commercial, navigational, industrial and military skills based on measurement and mathematics. For mentality, nothing less than a new model of reality is required – a quantitative model. Crosby's thesis and arguments depend heavily on the development of practical mathematics.[38]

5. Conclusion

All history is about the present.[39] Thus, scientists are likely continually to find new questions about science that require historical investigation, which may in turn lead to something that proves to be either historically trivial or, perhaps less often, of great significance. The history of computer science raises questions that lead to the close study of the development and dissemination of mathematics for practical application in the world's work. In particular, the development of practical mathematics in the Tudor period is found to be an essential part of the history of computer science. In the British Isles, the aims and methods of practical mathematics were established by Recorde. Already detached from the academic mathematical tradition in the Tudor period, Recorde's work has been neglected in the historiography of mathematics.

Tudor mathematical practice is tightly bound to the role of quantification in society: goods, money, time and position. The merchant economy, broadly conceived, drove the development of quantification and created an important role in society for data and computation, underpinned by mathematical education. A mentality of quantification was influential in the extraordinary development and dominance of European society.

Robert Recorde lived at the centre of change in Britain, engaging politically, economically, socially and, of course, intellectually with these changes. To understand and appreciate Recorde better, we must place his work at the centre of a large European canvas. A sustained programme of research, publication and dissemination is a necessary, worthy and long overdue memorial to his achievements.

Notes

In preparing this chapter, I am grateful to Prys Morgan, Nia Powell, Ulrich Reich, Leo Rogers and Jack Williams for their comments and suggestions. I am particularly indebted to Adam Mosley, Tracey Rihll and Gareth Roberts for their careful and critical reading of earlier drafts.

1   The phrase 'May you live in interesting times' is the first of three supposed Chinese curses, the second being 'May you be known to important people'. In his dealings with Sir William Herbert, Recorde was twice cursed. The curses appear to have originated in America in the 1930s.

2   The theme of data has a central but somewhat neglected role in the history of computing, whose focus has tended to be on either the technicalities of software and hardware or on businesses and markets.

3   The analysis of the history of many technical subjects, other than those within science and technology, is dependent on the expertise of practitioners. For example, financial practitioners have analysed the development of banking in the ancient world (see Edward E. Cohen, *Athenian Economy and Society: A Banking Perspective* (Princeton, 1992), and David Jones, *The Bankers of Puteoli: Finance, Trade and Industry in the Roman World* (Stroud, 2006)). Engineering history can be dominated by the technical agendas of practitioners (see David Cannadine, 'Engineering history, or the history of engineering? Re-writing the technological past', *Transactions of the Newcomen Society*, 74/1 (2002–3), 163–80).

4   Given the scale of alchemy, the distinction of its practitioners (including Newton) and the obvious fact that chemistry emerged from centuries of systematic research on materials, there are plenty of reasons to tackle the subject. However, philosophical conceptions of science have long denied the relevance of alchemy. Of course, there were other reasons for scholars' antipathy, such as its inherent difficulty, the pressure of other priorities and a lack of scholastic confidence in the field. Alchemy has now been admitted to the history of science. See, for example, Stanton J. Linden (ed.), *The Alchemy Reader* (Cambridge, 2003), and William R. Newman and Lawrence M. Principe, *Alchemy Tried in the Fire: Starkey, Boyle, and the Fate of Helmontian Chymistry* (Chicago, 2002).

5   I. Grattan-Guinness, 'Talepiece: the history of mathematics and its own history', in *idem* (ed.), *Companion Encyclopedia of the History and Philosophy of the Mathematical Sciences* (London, 1994), pp. 1665–75.

6   Historians of several kinds have preserved Recorde's memory, but I do not know the historiography. The remarkable scholar James Orchard Halliwell (1820–89) began his career with an interest in science and devoted his *The Connexion of Wales with the Early Science of England* (London, 1840) to Robert Recorde. Later Halliwell edited John Dee's diaries. See H. W. Dickinson, 'J. O. Halliwell and the Historical Society of London', *Isis*, 18 (1932), 127–32, and Theodore Hornberger, 'Halliwell-Phillipps and the history of science', *Huntington Library Quarterly*, 12/4 (1949), 391–9. Browsing histories of mathematics at school, I first met Recorde in W. W. Rouse Ball's *A Short Account of the History of Mathematics* (4th edn, London, 1919; New York, 1960), where his birth in Tenby was noted to my great interest. Recorde appears in most general histories of mathematics, sometimes with a facsimile from *The Whetstone of Witte*. However, appreciations of his *oeuvre* in specialized histories of algebra are limited and his work is not treated as a topic: see, for example, Jacqueline A. Stedall, *A Discourse Concerning Algebra: English Algebra to 1685* (Oxford, 2002). Recorde is the subject of the first chapter in Geoffrey Howson, *A History of Mathematics Education in England* (Cambridge, 1982), and he has not been forgotten by the

historians of money. His name is noted in the standard work by C. E. Challis, *The Tudor Coinage* (Manchester, 1978), where quotations from *The Ground of Artes* are used to make sense of the variety and relative values of the coinage in circulation in 1543 (pp. 221–2). L. V. Grinsell, *The Bristol Mint* (Bristol, 1972) also notes Recorde.

7   Turing's famous paper was published in 1936: A. M. Turing, 'On computable numbers, with an application to the Entscheidungsproblem', *Proceedings of the London Mathematical Society*, 2nd ser., 42 (1936), 230–65. Here we find a fundamental model of a computer, a proof that there are computers that are universal (some computers can be programmed to simulate any other computer) and an example of a problem that no computer can solve.

8   For example, the highly influential *Lisbon Agenda* to make the EU 'the most dynamic and competitive knowledge-based economy in the world capable of sustainable economic growth with more and better jobs and greater social cohesion, and respect for the environment by 2010' was adopted by the European Council in 2000.

9   Peter F. Drucker, *The Age of Discontinuity: Guidelines to Our Changing Society* (New York, 1969).

10   ESRC web page, downloaded July 2008, no longer available. The definition struck a chord with several commentators on policy, see Ian Brinkley, *Defining the Knowledge Economy*, Research Report for the Work Foundation (London, 2006).

11   This idea is commonly attributed to R. De Roover, 'The commercial revolution of the thirteenth century', *Bulletin of the Business Historical Society*, 16/2 (1942), 34–8.

12   L. E. Sigler, *Fibonacci's Liber Abaci: A Translation into Modern English of Leonardo Pisano's Book of Calculation* (New York, 2002).

13   For an excellent short introduction, see Warren Van Egmond, 'Abbacus arithmetic', in I. Grattan-Guinness (ed.), *Companion Encyclopedia*, pp. 200–9. There is an indispensable scholarly account of the abbacus tradition in Warren Van Egmond, *Practical Mathematics in the Italian Renaissance: A Catalog of Italian Abbacus Manuscripts and Printed Books to 1600* (Florence, 1980). To study an original abbacus arithmetic is more difficult: see D. E. Smith's translation of the Treviso Arithmetic of 1478, which is included in Frank J. Swetz, *Capitalism and Arithmetic: The New Math of the Fifteenth Century* (La Salle, Ill., 1987) and, more recently, Jens Høyrup, *Jacopo da Firenze's Tractatus Algorismi and Early Italian Abbacus Culture* (Basel, 2007). Although the mathematics is simple, the scholarship is difficult.

14   See Stephen Chrisomalis, *Numerical Notation: A Comparative History* (New York, 2010).

15   By the 1530s there were many printed German arithmetics by authors including Johannes Widmann (*c.*1462–98), Jacob Köbel (1470–1533), Michael Stifel (1487–1567), Adam Riese (1492–1559) and Christoff Rudolff (1499–1545). In Widmann's *Behende vnd hubsche Rechenung auff allen Kauffmanschafft* (Leipzig, 1489), the symbols + and − first appeared. Riese published *Rechenung auff der Linihen und Federn* (Erfurt, 1522), covering the calculating board and numerical calculations with Hindu–Arabic digits for businessmen and craftsmen, which ran to over a hundred editions. The compendium of Gregor Reisch (*c.*1467–1525), *Margarita philosophica* (Freiburg, 1503), is known for its splendid images including that of a female personification of arithmetic viewing a

calculation using Hindu–Arabic numerals in competition with a calculation performed on an abacus.

16   Salomon Bochner, 'Why mathematics grows', *Journal of the History of Ideas*, 26/1 (1965), 3–24.

17   See Petrus Apianus, *Eyn newe unnd wohlgegründte Underweysung aller Kauffmanns Rechnung* (*A New and Reliable Instruction Book of Calculation for Merchants*) (Ingolstadt, 1527). The painting was acquired by the National Gallery in 1890 and stimulated a great deal of research. The page of the book is sufficiently visible to have allowed W. F. Dickes to identify the commercial arithmetic in 1892. See John North, *The Ambassadors' Secret: Holbein and the World of the Renaissance* (London, 2002), where the page is reproduced (p. 159).

18   Again, see John North, *The Ambassadors' Secret* or S. Foister, A. Roy and M. Wyld, *Making and Meaning: Holbein's 'Ambassadors'*, National Gallery exhibition catalogue (London, 1997).

19   See Gerard L'E. Turner, *Elizabethan Instrument Makers: The Origins of the London Trade in Precision Instrument Making* (Oxford, 2000).

20   One thinks of the building boom in London such as in The Strand where Protector Somerset was building the lavish Somerset House amid much controversy.

21   See chapter 5.

22   Leonard Digges, *A Geometrical Practise Named Pantometria* (London, 1571).

23   For a recent account, see Deborah E. Harkness, *The Jewel House: Elizabethan London and the Scientific Revolution* (New Haven, 2007).

24   The development of professional and mathematical expertise is discussed in Eric H. Ash, *Power, Knowledge and Expertise in Elizabethan England* (Baltimore, 2004).

25   For an account of Sharrington, see Challis, *The Tudor Coinage*, pp. 101–3.

26   Jack Williams, 'Mathematics and the alloying of coinage 1202–1700: Part I', *Annals of Science*, 52/3 (1995), 213–34 and 'Mathematics and the alloying of coinage 1202–1700: Part II', ibid., 235–63.

27   The letter and trial are important sources of evidence for Recorde's history and are discussed in chapter 1. For a full account of Recorde's encounter with Herbert, see Jack Williams, *Robert Recorde: Tudor Polymath, Expositor and Practitioner of Computation* (London, 2011).

28   N. P. Sil, 'Sir William Herbert, Earl of Pembroke (c.1507–70): in search of a personality', *Welsh History Review*, 11/1 (1982), 92–107.

29   The problem is suggested by Jack Williams (see chapter 1). Linking the generations, there is some evidence that Bacon knew Dee and Dee knew Recorde. For example, an entry in John Dee's diary for 11 August 1582 probably refers to a visit by Bacon to Dee's home; see Edward Fenton (ed.), *The Diaries of John Dee* (Charlbury, 1998), p. 46. Dee had much in common with Recorde (e.g. the Muscovy Company, and Dee's later editorship of the *Ground* after Recorde's death). Intriguingly, Dee had good relations with Sir William Herbert; see R. Julian Roberts, 'Dee, John', *Oxford Dictionary of National Biography* (Oxford,

2004). Dee seems to have been well acquainted with most aspects of Elizabethan society. For relevant background on Bacon, see Eric H. Ash, *Power, Knowledge, and Expertise in Elizabethan England*, ch. 5.

30   Herbert Butterfield, *The Origins of Modern Science*, 1300–1800 (London, 1949).

31   For an early re-evaluation or reworking of the nature of the 'scientific revolution', see David C. Lindberg and Robert S. Westman (eds), *Reappraisals of the Scientific Revolution* (Cambridge, 1990). For an excellent critique of the idea that modern science emerged in the seventeenth century, see Andrew Cunningham and Perry Williams, 'De-centring the "big picture": *The Origins of Modern Science* and the modern origins of science', *British Journal for the History of Science*, 26 (1993), 407–32. Cunningham and Williams emphasize strongly that historical research should seek to understand what its actors thought of their enterprises. The notion of 'scientific revolution' belongs to the study of what historians thought of their enterprises. Cunningham and Williams are prepared to consider questions such as when we may consider science to have become 'modern'; indeed, they present a case for the 'age of revolutions' (the period 1760–1848). For a survey of literature on the 'scientific revolution', see H. Floris Cohen, *The Scientific Revolution: A Historiographical Inquiry* (Chicago, 1994).

32   Edgar Zilsel, 'The sociological roots of science', *American Journal of Sociology*, 47/4 (1942), 544–62. The article is reprinted along with otherwise unpublished material by Zilsel in a collection edited by Diederick Raven, Wolfgang Krohn and Robert S. Cohen, *The Social Origins of Modern Science* (Dordrecht, 2000).

33   Richard W. Hadden, *On the Shoulders of Merchants: Exchange and the Mathematical Conception of Nature in Early Modern Europe* (Albany, NY, 1994).

34   The quotation is a part of Recorde's comments on Ptolemy in *The Castle of Knowledge*. It is carved into the Robert Recorde Memorial at the Department of Computer Science, Swansea University. The Memorial was designed by John Howes and carved by Ieuan Rees in 2001.

35   Modernity is the subject of various social theories and is a highly contested matter. For the record, I have in mind the theory of modernity as developed in Anthony Giddens, *The Consequences of Modernity* (Cambridge, 1990). For Giddens, the key social attributes (see pp. 16–17) are (i) the separation of time and space, which he calls *time-space distanciation*; (ii) the disembedding of social systems from, and their re-embedding in, time-space; and (iii) the reflexive ordering and re-ordering of social relations. In particular, this disembedding and re-embedding are enabled by various *abstract systems*, which can be separated into two types called *symbolic tokens* and *expert systems*. The latter are 'systems of technical accomplishment or professional expertise that organise large areas of the material and social environments in which we live today' (p. 27). Examples are money and the system of banking.

36   For a historiographical analysis of these notions for the early modern period, see Garthine Walker, 'Modernization', in Garthine Walker (ed.), *Writing Early Modern History* (London, 2005), pp. 25–48.

37   Alfred W. Crosby, *The Measure of Reality: Quantification and Western Society, 1250–1600* (Cambridge, 1997).

38   Another example of an intellectual historian's approach to the development of an understanding of practical mathematics is provided by Mary Poovey's consideration of the epistemological challenge to understand the origins and nature of the modern concept of fact. In *A History of the Modern Fact: Problems of Knowledge in the Sciences of Wealth and Society* (Chicago, 1998) she argues that the mathematical and quantitative transformation of commerce over our period of interest established socially the idea of a fact as a funda-mental 'unit of knowledge and explanation'. This prepared the way for fact-dependent scientific and technological transformations that are validated empirically. Her analysis includes a detailed study of John Mellis's *A Briefe Instruction* (London, 1588).

39   The statement 'All history is about the present' is an aphorism that applies to a range of modern philosophical views on history; see E. H. Carr's influential *What is History?* (London, 1961). For a commonly cited early view, see Benedetto Croce, *History as the Story of Liberty* (London, 1941): 'The practical requirements which underlie every historical judgment give to all history the character of "contemporary history", because, however remote in time events there recounted may seem to be, the history in reality refers to present needs and present situations wherein those events vibrate' (p. 19).

From Recorde to relativity: a speculation[1]

GARETH WYN EVANS

IT WAS MY VERY GOOD FORTUNE, when I first went to university, to be sent to a college where practically everyone studied history. The result of this is two indelible impressions – the first the need to study history, the second the futility of studying history. For it was when I looked at historians and their ways that I think I started grasping what is probably true of all human thought in the last analysis, that to travel brings benefits untold, even though one never arrives, because to arrive is impossible. I put these considerations to you therefore as an uninhibited amateur, based on some considerable personal interest but on rather scanty and spasmodic reading. They are speculative and no more; they may very well be wrong.

I am going to be concerned with the problem of motion and with cosmology – which means, in this period, the problem of the motion of the planets. I take it as unnecessary here to disown the popular legend that the scientific revolution which ended with the synthesis of Newton was brought about because people started doing experiments, observing nature and avoiding hypothesis and speculation. Great achievements in human thoughts are not bought so cheaply, however much some popularizers of the so-called scientific method wish to debase the currency. In fact the whole 150 years from Copernicus and Recorde to Newton is shot through and through with abstract speculation. One will never begin to appreciate this whole achievement if one insists on regarding the Newtonian synthesis as something obvious and immediate which humans had been particularly obstinate and *twp* (stupid) not to have grasped before. In fact, the Newtonian synthesis bristles with difficulties and was indeed regarded at the time by most of the continental mathematicians as reactionary and obscurantist. All I want at this stage to insist upon is that we need a prior regard for it all as a mighty effort, long in the making, and that we look at the threads of thought as they gather in the age of Recorde to converge upon their focus a hundred years later in time.

When one tries to do this, to isolate ideas, at any time in the Renaissance, one meets a particular and characteristic difficulty. Our specialisms in knowledge did not then exist: all these thinkers are operating in what for us are different distinct fields, but for them they were all one. Newton was profoundly interested in cosmology, in alchemy, possibly even in astrology. We think of Christopher Wren as an architect – he was in fact a professor of geometry when he was thirty and had a deep knowledge of anatomy. Neither should we underestimate the calibre and stature of this knowledge. Wren, for example, performed two vital experiments in vivisection – his knowledge had to stand the test that the animal should survive the operation, for what he was after could not be established unless the animal should indeed survive. He was the first to inject directly into the blood stream – he injected some opium into a vein in a dog's leg and measured the time the opium took to produce its soporific effect and the time that it took the dog to recover. His removal of a dog's spleen was even more remarkable. The spleen is an organ deep in the abdomen, nestling under the ribs on the left of the body. Anatomists had always been puzzled by its function – indeed I think they still are. Wren wondered whether an animal could live without it, removed it surgically in a living dog and proved his point, for the dog recovered and lived on apparently none the worse. The point I wish to make is that the knowledge and dissecting skill behind this must have been of the first order – even now in an age of anaesthetics and asepsis this operation, which is sometimes performed on humans, is a major operation.

As we go back towards the beginning of this period, and to the time of Recorde in particular, this universally wide canvas of an individual's thought is probably a factor of increasing importance. We remember Recorde as a mathematician. He was also an astronomer – one of the first to accept Copernicus' revolution in cosmological thought. He lived in a time of high and dangerous theology, his mature years were spent in the periods of Reformation and Counter Reformation of Edward and Mary Tudor. He must have been a leading physician, for kings and queens always command the best in this line. And although his visit to prison was a debtor's and not a martyr's, and the health of Edward and Mary was no real recommendation for his medicine, yet the background is necessarily there.

All this means that one must be particularly on one's guard. If one is trying to trace the development of an idea in natural science, for example, it may well be that the real field of struggle and speculation, which contains the germ and even the development of that same idea before it becomes articulate in science, is some field that we in the twentieth century would consider unrelated. We run the risk, on the one hand, by focusing on what appears to us to be one particular activity of missing the real strategic movement. On the other, if we look at random, we run the risk of picking out causal connections that are phoney.

Therefore I propose to develop an argument in two stages. Firstly, I shall examine the scientific position, the scientific awareness of the problem of the motion of the planets, from the kind of awareness that Copernicus was trying to give to the kind of answers that Newton ultimately provided. I shall not be concerned at all with details; rather one must focus attention upon the scientific terms of reference, if I may use the phrase, upon the type of concept used to resolve the problem and in terms of which the answer at any particular stage is given. For it becomes almost immediately obvious that what would be regarded as an adequate answer changes profoundly – one is not concerned so much with refinements and developments inside a certain scheme of thought but rather with grasping the other end of the stick altogether. Secondly, when one uses these changes outside the scientific picture and recognizes them for what they are, it may be profitable to cast out more widely for this source of origin.

If I may anticipate my argument, I will summarize thus. The change from Copernicus to Newton amounts primarily to replacing scientific explanation expressed in scientific form by an explanation expressed in terms of incremental changes in time. For Copernicus, a planet moves in a circle because a circle is the most harmonious, indeed the only completely symmetrical curve. That is for him the ultimate answer, needing no explanation. I call it a form principle. For Newton, if a planet moves in a circle, it does so not because the circle is a symmetrical curve but because at any instant of time the movement of the planet in the next small increment of time is determined by certain laws. It may be a consequence of these laws that as increment follows increment the completed path may be a circle. But one's eye is never on the completed path: it is on the incremental law. Between the principle of completed form and the principle of incremental law there is an immense gulf requiring, above all else, a far more mathematical, numerical handling of space. This, in my contention, is the essential development before us.

But perhaps I had better first explain what the problem of the planets is. If you look into the sky on a clear moonless night, what you will see with the naked eye is two distinct kinds of objects. First you will see the stars appearing as sharp distinct points of light with a slight flicker. These stars appear always in the same position relative to one another – for example the well known configuration known as the Great Bear or the Plough is always recognizable as such and has been so since the earliest recorded history. Because of this permanence of relative position they are called fixed stars. And while maintaining their position relative to one another the whole firmament moves across the heavens every night just as the sun moves by day. Secondly one sees the planets. These are far fewer in number – up to the time of Newton only five were known – Mercury, which is always close to the sun and can only be seen in the early evening or just before dawn, Venus, Mars, Jupiter and Saturn. Most of these look far brighter than a star – anything that looks outstandingly bright in the sky is certainly one of them. But the real point of interest right up to the time of Newton was their extraordinary motion. For they move from night to night against the background of the fixed stars without rhyme or reason – sometimes to left, sometimes to right, sometimes staying still for a while, sometimes turning back upon their courses and for this reason the Greeks called them planets, which means wanderers. Up to the middle of the seventeenth century the central problem for cosmology was the explanation of the motion of the little wanderers.

The pre-Copernicus solution was as follows. The earth is the centre of the universe and the heavenly bodies move around the earth in paths which must fundamentally be circles – simply because this is the basic principle. Any scientific explanation must look beyond the apparent complexity for an ultimate harmony and order – this is a proposition valid for Aristotle, valid for Newton, valid for Einstein – all that differs is what one accepts as one's ultimate criterion of this harmony. At this period, this criterion would have been a principle of perfect completed form, hence the circle in its harmonious symmetrical whole. But the planets are certainly not moving uniformly in a circle, they are observed to wander erratically. Faced with this, thinkers did what they always did in science. They looked for this principle at a second remove, admitting that at first sight the basic form and structure is hidden, but it must be there below the surface all the same. So that they tried to build up the complicated observed behaviour from simple circular motion.

Imagine a wheel rolling on a fixed wheel and consider the path in space of a material point on the rim of the rolling wheel. It will look something like that in Figure 1 – or, more likely, that in Figure 2.

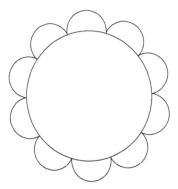

Figure 1. The locus of a point on a wheel rolling around a larger fixed wheel

Figure 2. A more likely depiction of the locus of a point on the rolling wheel

Yet this complicated curve is compounded of two simple circular motions. In such a manner cosmology before Copernicus tried to explain and predict the motion of the planets by building them up from perfect circular forms, with some success and immense complication.

Now please follow very carefully the change brought about by Copernicus in 1542. All this complication, he said, was quite unnecessary – the basic circular form was far nearer the surface of things than had been thought. The planets are moving in fact in perfectly straightforward circles uniformly and with perfect symmetry, but they move around the sun not around the earth. The complexity we see is due to the fact that we are looking from the earth, not from the sun, and is brought about because the earth is itself moving around the sun, also in that universal signature of all creation, the circle.

Now was not this all that had been looked for? With one bold stroke, the basic principle of circular form was established in its primitive purity so that a child could see. Was not this the final clear answer? Why then did Robert Recorde, according to the *Dictionary of National Biography*, although he accepted Copernicus' thesis, do so furtively, believing that the world was not yet ready for it? This qualification more than anything else that I have read about Recorde makes me think he was a thinker of the first magnitude. One will never really understand the ferment of mind in this field of natural philosophy in the next 150 years unless one grasps the implications of this change brought about by this deceptively simple step of Copernicus. It was not for nothing that almost half a century later the Church of Rome hounded Galileo and it was hardly the crystal clarity of new understanding that made Galileo support this doctrine of a moving earth. After all, he thought the existence of tides in the ocean proof positive of the motion that Copernicus had postulated, a standpoint which, I need hardly remind you, is quite irrelevant.

I have said already, and now say again, that every scientific synthesis is based necessarily upon a certain substratum of accepted principle, which is certainly not directly demonstrated. These principles are accepted as primary principles, and it is the business of scientific theory to resolve the whole manifest complexity of the universe around us to them. Before Copernicus these were principles of harmony of completed form.

It may happen that, at any time, a situation arises that calls in question not the competence of the scientific analysis, its technical evaluation, but something far more deep, nothing less than these basic first principles themselves. For example, almost in our own day, in the last eighty years, the Newtonian synthesis collapsed before the famous Michelson–Morley experiment which was again concerned with the motion of the earth, and later the basic principles of classical electromagnetic theory collapsed before what was called the Rayleigh–Jeans catastrophe. And both collapses generated a new synthesis, philosophically at least, relativity theory and quantum theory respectively. The Copernicus doctrine involved precisely such a collapse as Recorde seems to have appreciated. Although paradoxically it seemed at first sight to vindicate the basic principle, at a deeper level it swept it all away. For it cast into the melting pot the whole established explanation of gravitation.

Gravitation is the tendency of bodies to fall to the ground – one of the first things about the natural world every one of us observes. We explain this since we have been carefully taught to follow Newton by saying that the earth exerts a force upon bodies pulling them to itself – and by force we mean something similar to what my hand does to this table when I push it. Don't think that this is obvious or even right, for according to Einstein, it is all wrong. According to him, the force is precisely zero – there is no force. The effect is brought about because the earth changes the motion of space and time in its locality. I say this just to caution against a slap-happy attitude to the pre-Copernicus explanation that is before us.

Gravitation was explained by a principle similar I think in kind to the circular principle for the heavenly motions. They said: every particle of gross matter has its natural place to which it tends and that place is the centre of the universe. So that the earth does really nothing to the falling particle – the earth is really only the accumulation of matter particles which have already got as near as they can to this proper place. The particle is falling not to the earth but to the centre of the universe. So that it was essential for the whole explanation that the earth should remain at the centre of the cosmos – it was there not for any anthropomorphic prejudices but to comprehend gravitation.

In making the earth move, Copernicus destroyed all this, destroyed essentially Aristotelian physics in its entirety, for if the earth moved, it could not always be at this still centre. The theory fell because, in the new context, it could not handle gravitation at all. It was the end of an era – what we will not see again is the kind of principle that would have been accepted as an ultimate explanation, a geometrical form principle.

I want now to take up the story when scientific theory emerges again from the melting pot as a coherent whole with the publication of Newton's *Principia* in the 1680s. What I want to note is the different kind of answer now given. There is no looking for harmony of form in any geometrical path. The fact that the planets are moving nearly in circles around the sun simplifies nothing – one would not here be in the least disturbed if they were moving in paths as erratic as the sparks of a Catherine wheel. In fact, and this is the crucial point, one is not here concerned with the motion as a whole but rather with something quite different in kind. What one is concentrating on is not the wholeness of the motion, exhibited in the completeness of a geometrical path like the circle, but on something quite different. And that something is how the positions of the planets at successive intervals of time are related to one another. The ultimate reality, the ultimate structure of the whole universe is to be found not in any complete form but in the nature of the development of the system in time.

All this involves the growth of new concepts and of two in particular. It involves obviously a different awareness of the time process, and of handling it. It involves also, equally important, a different concept of space, the use of the vacuum. I need hardly remind you that right up to this time the idea that a vacuum existed and was a vital key to the natural world was anathema to most minds – something inconceivable. Nature abhors a vacuum.

The Newtonian answer to the motion of the planets is this: each planet moves in a vacuum. The path of each is due to a certain determinate development in time brought about by what is called the force of gravitation. This latter is something exceedingly strange – it is assumed to be similar to what we know as force – something brought about by strings and chains and brute muscle. But the force acting on the planets is presumably exerted by the vacuum which is emptiness, nothing – a hard and difficult position to maintain. Of course every subsidiary student swallows it without difficulty, an example of brainwashing if you like. But the essential of the Newtonian system is his exact handling of space and time and it is at these points that I want particularly to look.

Newton reports time as something universal, the same for all, running uniformly on. Nothing can hasten or retard its progress. It can be measured and that is all. Newton considers time as a river, something flowing inexorably on – something that depends on nothing else, a true first principle of the existence of the cosmos. And as time flows along, the inherent laws of nature determine how the whole of the natural world behaves and develops.

> Er pryder a gofalon
> Y mae ein heinioes ni
> Fel deilen ar yr afon
> Yn dilyn cwrs y lli.

— a verse that may be freely translated as

> 'Spite worries and trials
> Our life can but go
> As a leaf on a river
> That follows the flow.

All that is involved by implication, at least superficially, in Newton's handling of time.

I do not know where this concept of time originated as something, independent of all else, flowing uniformly on. I suspect that it was not a concept clearly held in the ancient world but I am going to venture the opinion that it was essentially well established in the Middle Ages. I have this opinion on a quite remarkable definition of time in *The Cloud of Unknowing*, a work of an unknown English mystic almost certainly of the second half of the fourteenth century. I have never seen it referred to. In the fourth chapter, we read:[2]

> This work asketh no long time or it be once truly done, as some men ween; for it is the shortest work of all that man may imagine. It is never longer, nor shorter, than is an atom: the which atom, by the definition of true philosophers in the science of astronomy, is the least part of time. And it is so little that for the littleness of it, it is indivisible and nearly incomprehensible. This is that time of the which it is written: All time that is given to thee, it shall be asked of thee, how thou hast dispended it. And reasonable thing it is that thou give account of it: for it is neither longer nor shorter, but even according to one only stirring that is within the principal working might of thy soul, the which is thy will. For even so many willings or desirings, and no more nor no fewer, may be and are in one hour in thy will, as are atoms in one hour.

It seems to me that this is precisely Newton's river of time. For the writer regards time as an infinite sequence of identical units, which he calls atoms, following one another. It differs only in that this time is not infinitely subdivisible, it has a least part. But provided that part is small enough, the concept converges into Newton's. I might also remark in passing that this issue of a least time, whether a smallest time exists, is a somewhat important issue in modern physics, but that of course is by the way. But the concept of vacuum, on the other hand, is certainly not present in the Middle Ages. It is almost entirely absent, and it must certainly be reported as one of the most important conceptual developments of this period. I propose to examine it in some detail.

The geometry of the Greeks was one of the finest developed mathematical disciplines of the ancient world and it has persisted up to our time as one of the greatest creations of the human mind. In handling a problem by Euclidean geometry one thinks of figure, triangles, parallelograms, circles – abstractions of certain essential material configurations. One is not concentrating at any stage on the vacuum in which these configurations are immersed.

Right in the middle of the period with which I am concerned Descartes invented what is called Cartesian geometry. It implied a subtly different way of handling geometry, indeed of handling space. For in Cartesian geometry one's eye is not on the geometrical forms – it is more upon the space in which they are immersed. To illustrate, consider the proof by the two methods of a simple theorem, say the opposite angles of a parallelogram are equal (see Figure 3).

Figure 3. To prove that the opposite angles of the parallelogram *ABCD* are equal

In Euclidean geometry, one proves that $B = D$ as follows. One breaks the figure up conceptually into two triangles, *ABC* and *ADC*. Then one shows that the two triangles are the same in the sense that the figure *ABC* could be moved about without distorting it so that it would lie exactly on top of *ADC*, just covering it completely, and with the angle *B* corresponding to the angle *D*. QED.

Now consider the same proof in Descartes' system. At no stage is one thinking of the parallelogram as a completed figure or breaking it up into simpler forms. One considers a straight line of infinite length of which *AB* is a part, considering it as a path in the space in which *A*, *B*, *C*, *D* are placed. Then one considers the path of which *BC* is a part and calculates the angle at which the two paths cross which is the angle *B*. One does the same thing for the angle *D* and finds that they are the same.

The difference in one's concepts in handling the problem is great. For Euclid, one's eye is on certain basic forms; for Descartes, one's eye is on an infinite sheet of graph paper.

The shift of emphasis that one finds here from some basic forms on the one hand to the containing nothingness on the other corresponds somewhat closely, I think, to the shift from the basic scientific handling of the problem of motion in the change from the Aristotelian system to the Newtonian. This changing awareness of space seems to me to be the most important single scientific ingredient of the whole period, and I want now finally to suggest its source. I shall suggest that the strategic steps were being taken in the handling of quite a different problem well before Newton, well before Descartes, in the first half of the sixteenth century, which is of course the age of Recorde himself.

I should perhaps warn you that I am now trying to handle a background of which I have very little knowledge. I have not read any of the works from which I shall quote – my source is *Architectural Principles in the Age of Humanism* by Rudolf Wittkower.[3] Therefore we move to the problem of how to make a building beautiful as it appeared to the architects of the Renaissance in Italy and consider the change of emphasis, of the basic principles of design as they were formulated from the time of Alberti who wrote about 1450 to the time of Palladio who wrote and built a hundred years later. In England, this period in building was somewhat dead, producing more highlights of the details of craft like fan-vaulting, rather than any particular principle of basic structure. The influence of the Italian movement in actual works of brick and stone is little felt until the time of Inigo Jones.

The first thing one notices most strongly is the great similarity in cast of mind between these thinkers and the thinkers with whom we have been concerned in science. First there is this immense dependence on classical authority and, in just the same way as the students of physics were for ever referring to Aristotle and taking their basic principles from him, so also these thinkers were referring to Vitruvius who wrote on architecture about the time of Christ. For the whole intellectual activity of the Renaissance the ancient world was always the starting point.

In his book on temples, Vitruvius had maintained that the proportions of the human figure should be reflected in the proportions of the temple. And a well-built man with arms extended, so he maintained again, fits exactly into a circle and a square, the most perfect geometrical forms. Therefore these forms should be the focus on which a building is based and from which the proportions of all things in the world can be defined. Thus, for example, Luca Pacioli, a mathematician and a friend of Leonardo da Vinci:[4]

First we shall talk of the proportions of man because from the human body derive all measures and their denominations and in it is to be found all and every ratio and proportion by which God reveals the innermost secrets of nature . . . For in the human body [the ancients] found the two main figures without which it is impossible to achieve anything, namely the perfect circle and the square.

And again Serlio, an architect of great influence:[5] 'I begin with the circular form because it is more perfect than all the others.' For them all, the circle and, to a lesser extent, the square have an almost magical power to evoke beauty. And their building activity reflects this theoretical position. The whole treatment is to enhance the geometrical scheme, based upon perfect circle and square.

So for these people one can quite clearly isolate a cast of mind almost identical with the early physicists, a cast of mind dominated by the perfect form and obviously an awareness of space and geometry dependent upon completed forms. But in the second half of the period, the early sixteenth century, a new ordaining idea is taking shape. Its importance cannot be overestimated. It is based upon a very old discovery of Pythagoras.

Reflect that these people were trying to define beauty, which they regarded, not as something subjective and variable, but as an objective reality, as much a part of the natural world, and as much a hidden secret, as the nature of the motion of the planets. Therefore their aim was to set out a scheme which would achieve beauty when carried out. The insistence on circle and square was one such scheme. But Pythagoras seemed to have discovered an immediate and clear connection. He had found that if two similar taut strings are plucked to give musical notes, two strings will produce together a melodious and harmonious tone that one will recognise as producing an effect of beauty only if their lengths are in the ratio of two small integers and, if their ratios are not so related, their sound will be harsh and unpleasant. The effect was clear – it led to a shift of emphasis, to the belief that the key to beauty lay not so much in form structures like circles but in numerical patterns. It is my belief that it was in the translation of this belief into practice, into brick and stone, that the new concentration upon space itself and not upon forms immersed in it took gradual shape.

For if you are designing a church on the principle of geometrical forms you take a piece of paper and draw upon it your arrangement of circles and squares. It is on these that you have your eye – the piece of paper is no more than a convenient container. So was the vacuum to the early physicists. But if your church has to be designed to have a numerical pattern – if the length and breadth of nave and chancel are to be determined not by the intersection of squares and circles but by numbers, your eye must necessarily be in the first place on the piece of paper itself. You have to mark it out as a graph paper is marked, you are interested not in forms but in distances in space itself. In fact you have moved from a handling of geometry by the method of Euclid to the method of Descartes.

The analogy of this whole process with the essential shift of emphasis in physics from Copernicus to Newton seems to me to be too strong to be coincidence. It seems to me that one is face to face with a strategic movement in human thought. To it I would refer the attempt always to formulate problems in a mathematical numerical fashion in the sixteenth century and I do not think that it is extravagant to suggest that Recorde as a pioneer in this movement was caught in the same stream.

It may also be salutary to reflect that in this age of ours, when science in its deeper significance, in its philosophic role, is fighting a losing battle with a mechanical and brute technology, a decisive step in the truth of modern science was generated in the quest for beauty, for the harmony, even the poetry, of creation.

July 1958

Notes

1   The 400th anniversary of the death of Robert Recorde was commemorated in 1958 at a conference in Carmarthen. A report on the conference written by W. S. Evans, Honorary Secretary of the south-west Wales branch of the Mathematical Association, appeared in the *Mathematical Gazette*, 43/343 (February 1959), i–ii. It appears that none of the papers presented at this conference was published. One of the lecturers was Gareth Wyn Evans, who died in 2005. In that year Catrin Thomas, Gareth's daughter and herself a mathematician, was sorting out her father's papers. To her great surprise she discovered his notes for the lecture, written out in detail in an old University of Wales examination book! These notes are reprinted here as a period piece with minimal editing and with the kind permission of the Estate of Gareth Wyn Evans. They were printed by WJEC as a limited-copy booklet in 2008 to coincide with the conference held to commemorate the 450th anniversary of the death of Robert Recorde.

2   *The Cloud of Unknowing*, ed. Evelyn Underhill (London, 1922) at *http://www.sacred-texts.com/chr/cou /cou09.htm*, accessed 21.05.10.

3   Rudolf Wittkower, *Architectural Principles in the Age of Humanism*, Studies of the Warburg Institute, 19 (London, 1949). Quotations are taken from the third edition (London, 1962).

4   Ibid., p. 15.

5   Ibid., p. 18.

Bibliography

Works by Robert Recorde

We reproduce here some of the detail that appears on the title pages of each of Recorde's extant books, together with information about the printer that appears on the title page, in the colophon or in both places in each book. In the case of *The Ground of Artes*, information about both the first edition (1543) and Recorde's revised 1552 edition is included.

The Ground of Artes, R. Wolfe, London 1543.

> *The groūd of artes teachyng the worke and practise of Arithmetike, moch necessary for all states of men. After a more easyer & exacter sorte, then any lyke hath hytherto ben set forth: with dyuers newe additions, as by the table doth partly appeare.* ROBERT RECORDE.

> *Imprynted at London in Powls church yarde at the sygne of the Brasen serpent by R. Wolfe. In the yeare of our Lord Christ M. D. xliii. in October*

The Ground of Artes, Reynold Wolff, London 1552.

> THE GROVND OF ARTES *Teachyng the worke and practise of Arithmetike. Bothe in whole numbres and fractions, After a more easyer and exacter sorte, than any lyke hath hytherto been sette forth: with diuers new additions, as by the table doeth partely appeere. Made by M.* ROBERT RECORDE *Doctor of Physike.* 8. Octobre. 1552. *Imprinted at London, by Reynold Wolff.*

Other editions: 1545, 1549, 1551, 1558, 1561, 1566, 1570, 1571, 1573, 1575, 1579, 1582, 1594, 1596, 1607, 1610, 1615, 1618, 1623, 1631, 1632, 1636, 1640, 1646, 1648, 1652, 1654, 1658, 1662, 1668, 1673, 1699. Facsimile of the 1543 first edition, TGR Renascent Books, Derby 2009.

(See Joy B. Easton, 'The early editions of Robert Recorde's *Ground of Artes*', *Isis*, 58/4 (1967), 515–32.)

The Vrinal of Physick, Reynolde Wolfe, London 1547.

The Vrinal of Physick. Composed by Mayster ROBERT RECORDE: *Doctor of Physicke. 1547. Imprinted at London by Reynolde Wolfe.*

Other editions: 1548, 1558, 1559, 1567, 1574, 1582, 1599, 1651, 1665, 1679. Facsimile of the 1547 first edition, TGR Renascent Books, Derby 2011.

(See W. F. Sedgwick, 'Robert Recorde', in *Oxford Dictionary of National Biography* (Oxford, 1896), which may be accessed via the *DNB Archive* link in the 2009 ODNB online entry on Robert Recorde by Stephen Johnston. The title of the 1679 edition was changed to *The Judgement of Urines*, but is based on Recorde's original text.)

The Pathway to Knowledg, Reynold Wolfe, London 1551.

The pathway to KNOWLEDG CONTAINING THE FIRST PRIN*ciples of Geometrie, as they may moste aptly be applied vnto practise, bothe for vse of instrumentes Geometricall, and astronomicall and also for proiection of plattes in euerye kinde, and therfore much necessary for all sortes of men.*

IMPRINTED at London in Poules churcheyarde, at the signe of the Brasen serpent, by Reynold Wolf. *ANNO DOMINI M.D.L.I.*

Other editions: 1574, 1602. Facsimile of the 1551 first edition, TGR Renascent Books, Derby 2009.

The Castle of Knowledge, Reginalde Wolfe, London 1556.

THE CASTLE OF KNOWLEDGE CONTAINING THE EXPLICATION OF THE SPHERE *bothe celestiall and materiall, and diuers other thinges incident therto. With sundry pleasaunt proofes and certaine newe demonstrations not written before in any vulgare woorkes.*

Imprinted at London by Reginalde Wolfe, Anno Domini, 1556.

Other edition: 1596. Facsimile of the 1556 first edition, TGR Renascent Books, Derby 2009.

The Whetstone of Witte, Jhon Kyngstone, London 1557.

> *The whetstone of witte, whiche is the seconde parte of Arithmetike: containyng thextraction of Rootes: The Cossike practise, with the rule of Equation: and the woorkes of Surde Nombers.*

> *These Bookes are to bee solde, at the Weste doore of Poules, By Jhon Kyngstone.*

> *At London the .xii. daie of Nouember .1557.*

> *Imprinted at London, by Jhon Kyngston. Anno domini. 1557.*

No subsequent editions. Facsimile of the 1557 edition, TGR Renascent Books, Derby 2010.

In a project spanning a decade from the late 1960s to the mid 1970s, two reprint houses, Da Capo Press (New York) and Theatrum Orbis Terrarum Ltd (Amsterdam), combined their resources to make photographic facsimile copies of 964 texts published in England between 1475 and 1640. Under the general title *The English Experience: Its Record in Early Printed Books Published in Facsimile* the series includes editions of Recorde's works. The quality of the reproduction is variable and differs qualitatively from that achieved in the newly typeset facsimile copies published by TGR Renascent Books listed above.

There is evidence that Recorde wrote at least two other books, which are no longer extant, namely *The Gate of Knowledge* and *The Treasure of Knowledge*. Sedgwick ('Robert Recorde') suggests that the former was probably on mensuration and the latter probably on the 'higher part of astronomy'. He also sets out the basis for other speculations that Recorde may have written further works on mathematics, as well as on anatomy, British history, the natural sciences, navigation and theology. Recorde's literary achievements also included providing material for a revised fourth edition of Robert Fabyan's *Chronicle*, published in 1559.

Primary

Alingham, William, *An Epitome of Geometry, being a Compendious Collection of the 1st, 3rd, 5th, 6th, 11th and 12th Books of Euclid* (London, 1695).

Anon., *A Discourse of the Common Weal of this Realm of England*, ed. E. Lamond (Cambridge, 1929).

Anon., *An Introduction for to Lerne to Recken with the Pen and with Counters* (London, 1536/7, 1539). Facsimile of the 1539 edition, TGR Renascent Books (Derby, 2009).

Anon., *The Key to Unknowne Knowledge. Or, a Shop of five Windowes. . . Consisting of five necessarie Treatises: Namely, 1. The Judgement of Urines 2. Judiciall rules of Physike. . .* (London, 1599).

Apianus, Petrus, *Eyn newe unnd wohlgegründte Underweysung aller Kauffmanns Rechnung* (Ingolstadt, 1527).

Bale, John, *Index Britanniae scriptorum: John Bale's Index of British and Other Writers*, ed. R. L. Poole and M. Bateson; reissued with Introduction by C. Brett and J. P. Carley (Cambridge, 1990).

Barrow, Isaac, *Euclide's Elements: The Whole Fifteen Books Compendiously Demonstrated* (London, 1660).

Benese, Richard, *This Boke Sheweth the Maner of Measurynge of all Maner of Lande* (London, 1537).

Billingsley, Henry, *The Elements of Geometrie of the Most Auncient Philosopher Euclide* (London, 1570).

Boethius, *Hec sunt opera Boetii* (Venice, 1491).

Borough, William, *A Discovrs of the Variation of the Cumpas* (London, 1581).

Byrne, Oliver, *The First Six Books of the Elements of Euclid* (London, 1847).

Caius, John, *A Boke or Counseill against the Disease called the Sweate* (London, 1552; facsimile edited by Archibald Malloch, New York, 1937).

Capella, Martianus, *Opus Martiani Capellae* (first published 1491).

Cardano, Girolamo (or Hieronimo), *Practica arithmetice, & mensurandi singularis* (Milan, 1539).

Cardano, Girolamo (or Hieronimo), *Artis magnae, sive de regulis algebraicis* (Nuremberg, 1545).

Copernicus, Nicolaus, *De revolutionibus orbium caelestium* (Nuremberg, 1543).

Cuningham, William, *The Cosmographical Glasse* (London, 1559).

Dee, John, *The Mathematicall Praeface to the Elements of Geometrie of Euclid of Megara* (London, 1570: facsimile edition, New York, 1975).

Dee, John, *General and Rare Memorials Pertaynyng to the Perfect Arte of Navigation* (London, 1577).

Digges, Leonard, *A Geometrical Practise Named Pantometria* (London, 1571).

Dürer, Albrecht, *Underweysung der Messung* (Nuremberg, 1525).

Eden, Richard, *The Arte of Nauigation, Conteynyng a Compendious Description of the Sphere* (London, 1561); trans. from Martin Cortes, *Breve compendio de la sphera y de la arte de navegar* (1551).

Edward VI, King, *The Chronicle and Political Papers of King Edward VI*, ed. W. K. Jordan (London, 1966).

Elyot, Sir Thomas, *The Boke Named the Gouernour* (London, 1539).

Elyot, Sir Thomas, *The Castel of Helth* (London, 1539?).

Fabyan, Robert, *The New Chronicles of England and France in Two Parts*, ed. Henry Ellis (London, 1811).

Fine, Oronce (Orontius), *De mundi sphaera* (Paris, 1542).

Frisius, Gemma, *Arithmeticae practicae methodus facilis* (Antwerp, 1540).

Gairdner, James (ed.), *Letters and Papers, Foreign and Domestic, Henry VIII*, vol. 12, pt. 1 (London, 1890), pt. 2 (London, 1891).

Grynaeus, Simon, *Eukleidou Stoicheion bibl. 15* (Basel, 1533).

Halifax, William, *The Elements of Euclid Explain'd in a New, but Most Easie Method* (Oxford, 1685).

Hatton, Edward, *An Intire System of Arithmetic* (London, 1721).

Heath, Thomas Little, *The Thirteen Books of Euclid's Elements*, 3 vols (1908; reprinted New York: Dover, 1956).

Hooper, John, *A Declaration of the Ten Holy Comaundementes of Allmygthye God* (Zurich, 1548).

Huswirt, Johannes, *Enchiridion nouus algorismi summopere visus de integris* (Cologne, 1501).

Joyce, David, *Euclid's Elements* (2002) *aleph0.clarku.edu/~djoyce/java/elements/trip.html*

La Ramée, Pierre de (Petrus Ramus), *Remonstrance de Pierre de La Ramée faite au Conseil privé, en la Chambre du Roi au Louvre, le 18 janvier 1567, touchant la profession royale en mathematique* (Paris, 1567).

Lane, Joan (ed.), *John Hall and his Patients: The Medical Practice of Shakespeare's Son-in-Law* (Stratford-upon-Avon, 1996).

Leeke, John, and George Serle, *Euclid's Elements of Geometry in XV Books: With a Supplement of Divers Propositions and Corollaries* (London, 1661).

Lefèvre d'Étaples, Jacques, *Euclidis Megarensis geometricorum elementorum libri xv: Campani Galli Transalpini in eosdem commentariorum libri xv* (Paris, 1516; reprinted Basel, 1537).

Maxwell-Lyte, H. C. (ed.), *Calendar of the Patent Rolls Preserved in the Public Record Office* (London, 1891–1916).

Mellis, John, *A Briefe Instruction and Maner How to Keepe Bookes of Accompts after the Order of Debitor and Creditor* (London, 1588).

More, Sir Thomas, *A Fruteful and Pleasant Worke of the Beste State of a Publique Weale, and of the New Yle called Vtopia*, trans. Raphe Robynson (London, 1551).

Nicolas, Sir Harris, *Proceedings and Ordinances of the Privy Council of England, 1386–1542*, vol. 3: *1 Henry VI, 1422–7 Henry VI, 1429* (London, 1834).

Pacioli, Luca, *Summa de arithmetica, geometria, proportioni et proportionalità* (Venice, 1494).

Pacioli, Luca, *Euclidis Megarensis . . . opera* (Venice, 1509).

Prise, John, *Yny Lhyvyr Hwnn* (London, 1546).

Proclus, *De sphaera*, trans. Thomas Linacre (Venice, 1499).

Ratdolt, Erhard, *Preclarissimus liber elementorum Euclidis* (Venice, 1482).

Reisch, Gregor, *Margarita philosophica* (Freiburg, 1503).

Rheticus, Georg Joachim, *De lateribus et angulis triangulorum* (Wittenberg, 1542).

Riese, Adam, *Rechenung auff der Linihen und Federn* (Erfurt, 1522).

Rudd, Thomas, *Euclides Elements of Geometry: The First VI Books: In a Compendious Form Contracted and Demonstrated* (London, 1651).

Rudolff, Christoff, *Behend vnnd Hubsch Rechnung durch die kunstreichen regeln Algebre so gemeincklich die Coss genennt werden* (Strasbourg, 1525).

Rudolff, Christoff, *Künstliche Rechnung* (Vienna, 1526).

Sacrobosco, Johannes de, *Tractatus de sphaera* (c.1230, first printed Ferrara, 1472).

Sacrobosco, Johannes de, *Algorismus* (c.1250).

Salesbury, William, *The Descripcion of the Sphere or the Frame of the Worlde* (London, 1550).

Scheubel, Johann, *De numeris et diversis rationibus* (Leipzig, 1545).

Scheubel, Johann, *Algebrae compendiosa facilisque descriptio* (Paris, 1551/2).

Scheubel, Johann, *Das sibend, acht und neunt Buch Euclidis* (Augsburg, 1555).

Schreyber (Grammateus), Heinrich, *Ayn new kunstlich Buech* ... (1518, printed in Nuremberg, 1521).

Sigler, L. E., *Fibonacci's Liber Abaci: A Translation into Modern English of Leonardo Pisano's Book of Calculation* (New York, 2002).

Stifel, Michael, *Arithmetica integra* (Nuremberg, 1544).

Starkey, Thomas, *A Dialogue Between Pole and Lupset*, ed. T. F. Mayer, Camden Fourth Series, 37 (London, 1989).

Strype, John, *Ecclesiastical Memorials; Relating chiefly to Religion, and the Reformation of it*, 3 vols (London, 1721).

Tartaglia, Niccolò, *Euclide ... diligentemente reassettato* (Venice, 1543).

The Statutes of the Realm, vol. 3 (London, 1817).

Tunstall, Cuthbert, *De arte supputandi libri quattuor* (London, 1522).

Walkingame, Francis, *The Tutor's Assistant* (London, 1751).

Widmann, Johannes, *Behende vnd hubsche Rechenung auff allen Kauffmanschafft* (Leipzig, 1489).

Williams, Reeve, *The Elements of Euclid Explained and Demonstrated in a New and Easy Manner* (London, 1685).

Zamberti, Bartolomeo, *Euclidis Megarensis ... elementorum libros xiij* (Venice, 1505).

Secondary

Aalkjaer, V., 'Uroscopia: a historical and art historical essay', *Acta Chirurgica Scandinavica*, 433 (1973), 3–11.

Adams, Eleanor B., 'An English library at Trinidad, 1633', *The Americas*, 12/1 (1955), 25–41.

Alexander, Amir R., *Geometrical Landscapes: The Voyages of Discovery and the Transformation of Mathematical Practice* (Stanford, 2002).

Andrews, K. R., *Trade, Plunder and Settlement: Maritime Enterprise and the Genesis of the British Empire, 1480–1630* (Cambridge, 1984).

Arber, Agnes, *Herbals, their Origin and Evolution*, 3rd edn (Cambridge, 1986).

Archibald, R. C., 'The first translation of Euclid's *Elements* into English and its source', *American Mathematical Monthly*, 57 (1950), 443–52.

Ash, Eric H., *Power, Knowledge and Expertise in Elizabethan England* (Baltimore, 2004).

Baird, L. Y., 'The physician's "urynals and jurdones": urine and uroscopy in medieval medicine and literature', *Fifteenth-Century Studies*, 2 (1979), 1–8.

Baldwin, T. W., *William Shakspere's Small Latine and Lesse Greeke* (Urbana Ill., 1944).

Barany, Michael J., 'Translating Euclid's diagrams into English, 1551–1571', in Albrecht Heeffer and Maarten Van Dyck (eds), *Philosophical Aspects of Symbolic Reasoning in Early Modern Mathematics* (London, 2010), pp. 125–63.

Barnard, F. P., *The Casting-Counter and the Counting-Board: A Chapter in the History of Mathematics and Early Arithmetic* (Oxford, 1916).

Barrow-Green, June, '"Much necessary for all sortes of men": 450 years of Euclid's *Elements* in English', *BSHM Bulletin*, 21/1 (2006), 2–25.

Beier, Lucinda McCray, *Sufferers and Healers: The Experience of Illness in Seventeenth-Century England* (London, 1987).

Berkel, K. van, A. van Helden and L. Palm (eds), *A History of Science in the Netherlands: Survey, Themes and Reference* (Leiden, 1999).

Bloemendal, Jan and Chris Heesakkers (eds), *Bio-bibliografie van Nederlandse Humanisten* (*Biography and Bibliography of Dutch Humanists*), digital publication, DWC/Huygens Instituut KNAW (The Hague, 2009), *www.dwc.knaw.nl* (accessed 14.09.11).

Bochner, Salomon, 'Why mathematics grows', *Journal of the History of Ideas*, 26/1 (1965), 3–24.

Boxer, C. R., *The Portuguese Seaborne Empire, 1415–1825* (New York, 1970).

Boyer, C. J., *A History of Mathematics* (New York, 1991).

Brinkley, Ian, *Defining the Knowledge Economy*, Research Report for The Work Foundation (London, 2006).

Bruyère, N., *Méthode et dialectique dans l'œuvre de La Ramée: Renaissance et âge classique* (Paris, 1984).

Burnett, John, 'William Prout and the urinometer: some interpretations', in Robert G. W. Anderson, James A. Bennett and William F. Ryan (eds), *Making Instruments Count: Essays on Historical Scientific Instruments* (Aldershot, 1993), pp. 242–54.

Butterfield, Herbert, *The Origins of Modern Science, 1300–1800* (London, 1949).

Cannadine, David, 'Engineering history, or the history of engineering? Re-writing the technological past', *Transactions of the Newcomen Society*, 74/1 (2002–3), 163–80.

Carneiro de Mendoça, Barbara Heliodora, 'The influence of Gorboduc on King Lear', *Shakespeare Survey*, 13 (1960), 41–8.

Carr, A. D., *Medieval Anglesey* (Llangefni, 1982).

Carr, E. H., *What is History?* (London, 1961).

Cassels, J. W. S., 'Is this a Recorde?' *Mathematical Gazette*, 60/411 (March 1976), 59–61.

Challis, C. E., *The Tudor Coinage* (Manchester, 1978).

Chrisomalis, Stephen, *Numerical Notation: A Comparative History* (New York, 2010).

Clark, George, *A History of the Royal College of Physicians of London*, 2 vols (Oxford, 1964–6).

Cohen, Bertram, 'A tale of two paintings', *Annals of the Royal College of Surgeons of England*, 64 (1982), 4–12.

Cohen, Edward E., *Athenian Economy and Society: A Banking Perspective* (Princeton, 1992).

Cohen, H. Floris, *The Scientific Revolution: A Historiographical Inquiry* (Chicago, 1994).

Collinson, Richard, *The Three Voyages of Martin Frobisher*, Hakluyt Society (London, 1867).

Cook, H. J., *Matters of Exchange: Commerce, Medicine and Science in the Dutch Golden Age* (New Haven, 2007).

Copenhaver, B. P. and C. B. Schmitt, *Renaissance Philosophy* (Oxford, 1992).

Croce, Benedetto, *History as the Story of Liberty* (London, 1941).

Crosby, Alfred W., *The Measure of Reality: Quantification and Western Society, 1250–1600* (Cambridge, 1997).

Crossley, D. W., 'The management of a sixteenth-century ironworks', *Economic History Review*, 19/2 (1966), 273–88.

Cule, John, 'Some early hospitals in Wales and the Border', *National Library of Wales Journal*, 20/2 (1977), 97–130.

Cunningham, Andrew and Perry Williams, 'De-centring the "big picture": *The Origins of Modern Science* and the modern origins of science', *British Journal for the History of Science*, 26 (1993), 407–32.

Daniel, R. Iestyn (ed.), *Gwaith Ieuan ap Rhydderch* (Aberystwyth, 2003).

Daniel, R. Iestyn et al. (eds), *Cyfoeth y Testun: Ysgrifau ar Lenyddiaeth Gymraeg yr Oesoedd Canol* (Cardiff, 2003).

De Roover, R., 'The commercial revolution of the thirteenth century', *Bulletin of the Business Historical Society*, 16/2 (1942), 34–8.

Department of Education and Science, *Mathematics Counts: Report of the Committee of Inquiry into the Teaching of Mathematics in Schools under the Chairmanship of Dr W. H. Cockcroft* (London, 1982).

Dickinson, H. W., 'J. O. Halliwell and the Historical Society of London', *Isis*, 18 (1932), 127–32.

Diffie, Bailey W. and George D. Winnius, *Foundations of the Portuguese Empire 1415–1580*, I: *Europe and the World in the Age of Expansion* (Minneapolis, 1977).

Dimmock, S., 'Haverfordwest: an exemplar for the study of southern Welsh towns in the later Middle Ages', *Welsh History Review*, 22/1 (2004), 1–28.

Dimmock, S., 'Reassessing the towns of southern Wales in the later Middle Ages', *Urban History*, 32/1 (2005), 33–45.

Dimmock, S., 'Urban and commercial networks in the later Middle Ages: Chepstow, Severnside and the ports of southern Wales', *Archaeologia Cambrensis*, 152 (2005 for 2003), 53–68.

Dresser, Madge and Peter Fleming, *Bristol: Ethnic Minorities and the City 1000–2001* (London, 2007).

Drucker, Peter F., *The Age of Discontinuity: Guidelines to Our Changing Society* (New York, 1969).

Dwnn, Lewys, *Heraldic Visitations of Wales and Part of the Marches*, ed. Samuel Rush Meyrick (Llandovery, 1846).

Dyer, Alan D., 'The English sweating sickness of 1551: an epidemic anatomized', *Medical History*, 41 (1997), 362–84.

Dyer, Alan D. and David M. Palliser (eds), *The Diocesan Population Returns for 1563 and 1603* (Oxford, 2005).

Easton, Joy B., 'On the date of Robert Recorde's birth', *Isis*, 57/1 (1966), 121.

Easton, Joy B., 'The early editions of Robert Recorde's *Ground of Artes*', *Isis*, 58/4 (1967), 515–32.

Evans, W. S., 'S.W. Wales Mathematical Association: report for the session 1957–1958', *Mathematical Gazette*, 43/343 (February 1959), i–ii.

Elton, G. R., *Studies in Tudor and Stuart Government and Politics*, vol. III: *Papers and Reviews 1973–1981* (Cambridge, 1983).

Emden, A. B., *A Biographical Register of the University of Oxford AD 1501–1540* (Oxford, 1974).

Fee, Elizabeth and Theodore M. Brown (eds), *Making Medical History: The Life and Times of Henry E. Sigerist* (Baltimore, 1997).

Fenton, Edward (ed.), *The Diaries of John Dee* (Charlbury, 1998).

Fletcher, J. M., 'The Faculty of Arts', in J. McConica (ed.), *The History of the University of Oxford*, III: *The Collegiate University* (Oxford, 1986), pp. 157–99.

Foister, S., A. Roy and M. Wyld, *Making and Meaning: Holbein's 'Ambassadors'*, National Gallery exhibition catalogue (London, 1997).

Giddens, Anthony, *The Consequences of Modernity* (Cambridge, 1990).

Gliozzi, Mario, 'Cardano, Girolamo (1501–76)', in *Dictionary of Scientific Biography*, 3 (New York, 1971), pp. 64–7.

Grattan-Guinness, I., 'Talepiece: the history of mathematics and its own history', in *idem* (ed.), *Companion Encyclopedia of the History and Philosophy of the Mathematical Sciences* (London, 1994), pp. 1665–75.

Griffith, W. P., 'Schooling and society', in J. Gwynfor Jones (ed.), *Class, Community and Culture in Tudor Wales* (Cardiff, 1989), pp. 79–119.

Grinsell, L. V., *The Bristol Mint* (Bristol, 1972).

Griscom, Acton and Robert Ellis Jones (eds), *The Historia Regum Britanniae of Geoffrey of Monmouth* (London, 1929).

Gruffydd, R. Geraint, '*Yny Lhyvyr Hwnn* (1546): the earliest Welsh printed book', *Bulletin of the Board of Celtic Studies*, 23/2 (1969), 105–16.

Gruffydd, R. Geraint, 'Y print yn dwyn ffrwyth i'r Cymro: *Yny Lhyvyr Hwnn*, 1546', *Y Llyfr yng Nghymru = Welsh Book Studies*, 1 (1998), 1–20.

Gunther, R. T., *Early Science in Oxford*, I (Oxford, 1923).

Hadden, Richard W., *On the Shoulders of Merchants: Exchange and the Mathematical Conception of Nature in Early Modern Europe* (Albany, NY, 1994).

Hall, Marie Boas, *The Scientific Renaissance 1450–1630* (New York, 1962; republished 1994).

Halliwell, James Orchard, *The Connexion of Wales with the Early Science of England* (London, 1840).

Harkness, Deborah E., *The Jewel House: Elizabethan London and the Scientific Revolution* (New Haven, 2007).

Henry, B. W., 'John Dee, Humphrey Llwyd, and the name "British Empire"', *Huntington Library Quarterly*, 35 (1971–2), 189–90.

Hett, W. S., *Aristotle: Minor Works*, Loeb Classical Library, 307 (London and Cambridge, Mass., 1955 edn).

Hill, G. F., *The Development of Arabic Numerals in Europe* (Oxford, 1915).

Hoak, Dale, 'Rehabilitating the Duke of Northumberland: politics and political control 1549–53', in J. Loach and R. Tittler (eds), *The Mid-Tudor Polity, c.1540–1560* (London, 1980), pp. 29–51.

Hornberger, Theodore, 'Halliwell-Phillipps and the history of science', *Huntington Library Quarterly*, 12/4 (1949), 391–9.

Howson, Geoffrey, *A History of Mathematics Education in England* (Cambridge, 1982).

Høyrup, Jens, *Jacopo da Firenze's Tractatus Algorismi and Early Italian Abbacus Culture* (Basel, 2007).

Jardine, Nicholas, *The Birth of History and Philosophy of Science* (Cambridge, 1984).

Jardine, Nicholas, 'Epistemology of the sciences', in C. B. Schmitt and Q. Skinner (eds), *The Cambridge History of Renaissance Philosophy* (Cambridge, 1988), pp. 685–711.

Jayawardene, S. A., 'Luca Pacioli', in *Dictionary of Scientific Biography* (New York, 1971), 274–7.

Jayne, Sears, *Plato in Renaissance England* (Dordrecht, 1995).

Jenkins, R. T. et al. (eds), *The Dictionary of Welsh Biography down to 1940* (London, 1959).

Johnson, Francis R., *Astronomical Thought in Renaissance England: A Study of the English Scientific Writings from 1500 to 1645* (Baltimore, 1937).

Johnson, Francis R. and Sanford V. Larkey, 'Robert Recorde's mathematical teaching and the anti-Aristotelian movement', *Huntington Library Bulletin*, 7 (April 1935), 59–87.

Jones, David, *The Bankers of Puteoli: Finance, Trade and Industry in the Roman World* (Stroud, 2006).

Jones, J. Gwynfor (ed.), *Class, Community and Culture in Tudor Wales* (Cardiff, 1989).

Jones, T. Gwynn (ed.), *Gwaith Tudur Aled* (Cardiff, 1926).

Jones, Whitney R. D., *The Tudor Commonwealth, 1529–1559* (London, 1970).

Kaplan, Edward, 'Robert Recorde and the authorities of uroscopy', *Bulletin of the History of Medicine*, 37 (1963), 65–71.

Karpinski, L. C. (ed.), *Robert of Chester's Latin Translation of the Algebra of al-Khowarizmi* (New York, 1915).

Kassell, Lauren, 'Simon Forman's philosophy of medicine: medicine, astrology and alchemy in London, c.1580–1611' (unpublished DPhil thesis, Oxford University, 1997).

Kassell, Lauren, *Medicine and Magic in Elizabethan London. Simon Forman: Astrologer, Alchemist and Physician* (Oxford, 2005).

Katz, Victor J., *A History of Mathematics: An Introduction* (New York, 1993).

Ker, N. R., 'Sir John Prise', *The Library*, 5th ser., 10/1 (March 1955), 1–24.

Kiefer, Joseph H., 'Uroscopy: the artist's portrayal of the physician', *Bulletin of the New York Academy of Medicine*, 40 (1964), 759–66.

Kusukawa, S., *The Transformation of Natural Philosophy: The Case of Philip Melanchthon* (Cambridge, 1995).

Langley, Patricia, 'Why a pomegranate?' *British Medical Journal*, 321 (4 November 2000), 1153–4.

Leader, D. R., *A History of the University of Cambridge*, I: *The University to 1546* (Cambridge, 1989).

Leustra, Hendrik and Alex van den Brandhof (eds), *Biografisch Woordenboek van Nederlandse Wiskundigen* (*Biographical Dictionary of Dutch Mathematicians*), digital publication, DWC/Huygens Instituut KNAW (The Hague, 2009), *www.dwc.knaw.nl* (accessed 14.09.11).

Lewis, D. Gerwyn, *The University and the Colleges of Education in Wales 1925–78* (Cardiff, 1980).

Lilly, Samuel, 'Robert Recorde and the idea of progress: a hypothesis and verification', *Renaissance and Modern Studies*, 2 (1958), 3–37.

Lindberg, David C. and Robert S. Westman (eds), *Reappraisals of the Scientific Revolution* (Cambridge, 1990).

Linden, Stanton J. (ed.), *The Alchemy Reader* (Cambridge, 2003).

Lloyd, Howell A., *The Gentry of South West Wales, 1540–1640* (Cardiff, 1968).

Lloyd, Howell A., '"Famous in the field of number and measure": Robert Recorde, Renaissance mathematician', *Welsh History Review*, 20/2 (2000), 254–82.

Mack, Peter, 'The dialogue in English education in the sixteenth century', in M.-T. Jones-Davies (ed.), *Le Dialogue au temps de la Renaissance* (Paris, 1984), pp. 189–212.

McLean, Antonia, *Humanism and the Rise of Science in Tudor England* (London, 1972).

Maddison, Francis, Margaret Pelling and Charles Webster (eds), *Linacre Studies: Essays on the Life and Work of Thomas Linacre c.1460–1524* (Oxford, 1977).

Marr, Alexander (ed.), *The Worlds of Oronce Fine: Mathematics, Instruments and Print in Renaissance France* (Donington, 2009).

Mason, A. Stuart, 'The arms of the College', *Journal of the Royal College of Physicians*, 26 (1992), 231–2.

Miller, Genevieve (ed.), *A Bibliography of the Writings of Henry E. Sigerist* (Montreal, 1966).

Morley, Frank V., 'Finis coronat opus', *Scientific Monthly*, 10/3 (1920), 306–8.

Morley, H. T., 'Notes on Arabic numerals in medieval England', *Berkshire Archaeological Journal*, 50 (1947), 81–6.

Mosley, Adam, 'Objects of knowledge: mathematics and models in sixteenth-century cosmology and astronomy', in S. Kusukawa and I. Maclean (eds),

Transmitting Knowledge: Words, Images and Instruments in Early Modern Europe (Oxford, 2006), pp. 193–216.

Mosshammer, Alden A., *The Easter Computus and the Origins of the Christian Era* (Oxford, 2008).

Newitt, Malyn, *A History of Portuguese Overseas Expansion, 1400–1668* (London, 2005).

Newman, William R. and Lawrence M. Principe, *Alchemy Tried in the Fire: Starkey, Boyle, and the Fate of Helmontian Chymistry* (Chicago, 2002).

Nicholson, G., 'The Act of Appeals and the English Reformation', in C. Cross, D. Loades and J. Scarisbrick (eds), *Law and Government under the Tudors* (Cambridge, 1988), pp. 19–30.

North, John David, 'Astronomy and Mathematics', in J. I. Catto and T. A. R. Evans (eds), *The History of the University of Oxford*, II: *Late Medieval Oxford* (Oxford, 1992), pp. 103–74.

North, John David, *The Ambassadors' Secret: Holbein and the World of the Renaissance* (London, 2002).

Nutton, Vivian, 'John Caius and the Linacre tradition', *Medical History*, 23 (1979), 373–91.

Nutton, Vivian, *John Caius and the Manuscripts of Galen* (Cambridge: Cambridge Philological Society suppl. vol. 13, 1987).

Nutton, Vivian, 'Idle old trots, cobblers, and costermongers: Pieter van Foreest on quackery', in Henriette A. Bosman-Jelgersma (ed.), *Petrus Forestus Medicus* (Amsterdam, 1997), pp. 245–54.

Owen, Morfydd E., 'Prolegomena i astudiaeth lawn o lsgr. NLW 3026, Mostyn 88 a'i harwyddocâd', in R. Iestyn Daniel et al. (eds), *Cyfoeth y Testun: Ysgrifau ar Lenyddiaeth Gymraeg yr Oesoedd Canol* (Cardiff, 2003), pp. 349–84.

Oxford Dictionary of National Biography, 60 vols (Oxford, 2004), and online edn.

Parry, Graham, 'Patronage and the printing of learned works for the author', in John Barnard and D. F. McKenzie (eds), *The Cambridge History of the Book in Britain*, vol. IV: *1557–1695* (Cambridge, 2002), pp. 174–88.

Partington, J. R., *A History of Chemistry*, 4 vols (London, 1961–70).

Patterson, Louise Diehl, 'Recorde's cosmography, 1556', *Isis*, 42/3 (1951), 208–18.

Pedersen, Olaf, 'In quest of Sacrobosco', *Journal for the History of Astronomy*, 16 (1985), 175–220.

Pelling, Margaret, 'The women of the family? Speculations around early modern British physicians', *Social History of Medicine*, 8 (1995), 383–401.

Pelling, Margaret, 'Compromised by gender: the role of the male medical practitioner in early modern England', in Hilary Marland and Margaret Pelling (eds), *The Task of Healing: Medicine, Religion and Gender in England and the Netherlands 1450-1800* (Rotterdam, 1996), pp. 101–34.

Pelling, Margaret, *Medical Conflicts in Early Modern London: Patronage, Physicians, and Irregular Practitioners 1550–1640* (Oxford, 2003).

Pevsner, N., and E. Hubbard, *The Buildings of England: Cheshire* (Harmondsworth, 1971).

Piotti, Sonia, *The First Algebra Printed in English: The Whetstone of Witte (1557) of Robert Recorde* (Milan, 2005).

Pollard, A. F., *England under Protector Somerset: An Essay* (New York, 1966).

Poovey, Mary, *A History of the Modern Fact: Problems of Knowledge in the Sciences of Wealth and Society* (Chicago, 1998).

Powell, N. M. W., 'Robert ap Huw: a wanton minstrel of Anglesey', *Welsh Music History*, 3 (1999), 5–29.

Powell, N. M. W., 'Do numbers count? Towns in early modern Wales', *Urban History*, 32/1 (2005), 46–67.

Powell, N. M. W., '"Near the margin of existence"? Upland prosperity in Wales during the early modern period', *Studia Celtica*, 41/1 (2007), 137–62.

Power, D'Arcy, *William Harvey* (London, 1907).

Power, D'Arcy, *Notes on Early Portraits of John Banister, of William Harvey, and the Barber-Surgeons' Visceral Lecture in 1581* (London, 1912).

Pullan, J. M., *The History of the Abacus* (London, 1968).

Rankin, F. K. C., 'The arithmetic and algebra of Luca Pacioli (*c*.1445–1517)' (unpublished PhD thesis, Warburg Institute, University of London, 1992).

Raven, Diederick, Wolfgang Krohn and Robert S. Cohen (eds), *The Social Origins of Modern Science* (Dordrecht, 2000).

Reich, Karin, 'Michael Stifel', in Menso Folkerts, Eberhard Knobloch and Karin Reich (eds), *Maß, Zahl und Gewicht*, 2nd edn (Wiesbaden, 2001), pp. 66–89.

Reich, Ulrich, 'Scheubel(ius), Johann(es)', in *Neue Deutsche Biographie*, vol. 22, ed. Historische Kommission bei der Bayerischen Akademie der Wissenschaften (Berlin, 2005), pp. 709–10.

Roberts, Gordon, 'Robert Recorde: his life and times', *BSHM Bulletin*, 24/1 (2009), 40–2.

Roberts, Julian and Andrew G. Watson (eds), *John Dee's Library Catalogue* (London, 1990).

Rouse Ball, W. W., *A Short Account of the History of Mathematics*, 4th edn (London, 1919; New York, 1960).

Sherman, W. H., *John Dee: The Politics of Reading and Writing in the English Renaissance* (Amherst, 1995).

Sil, N. P., 'Sir William Herbert, Earl of Pembroke (*c*.1507–70): in search of a personality', *Welsh History Review*, 11/1 (1982), 92–107.

Siraisi, Nancy G., 'The music of pulse in the writings of Italian academic physicians (fourteenth and fifteenth centuries)', *Speculum*, 50 (1975), 689–710.

Siraisi, Nancy G., *Medieval and Early Renaissance Medicine: An Introduction to Knowledge and Practice* (Chicago, 1990).

Skeat, W. W. (ed.), *A Treatise on the Astrolabe addressed to his son Lowys by Geoffrey Chaucer A.D. 1391, edited from the Earliest MSS . . .*, Early English Texts Society, 16 (London, 1872).

Smith, David Eugene, *History of Mathematics*, vol. 1 (1923; reprinted New York: Dover, 1958), and vol. 2 (1925; reprinted New York: Dover, 1958).

Smith, David Eugene and Frances Marguerite Clarke, 'New light on Robert Recorde', *Isis*, 8/1 (1926), 50–70.

Stedall, Jacqueline A., *A Discourse Concerning Algebra: English Algebra to 1685* (Oxford, 2002).

Steele, Robert (ed.), *The Earliest Arithmetics in English* (London, 1922).

Strong, Roy C., 'Holbein's cartoon for the barber-surgeons group rediscovered – a preliminary report', *Burlington Magazine*, 105 (1963), 4–14.

Swetz, Frank J., *Capitalism and Arithmetic: The New Math of the Fifteenth Century* (La Salle Ill., 1987).

Tankard, Danae, 'Protestantism, the Johnson family and the 1551 sweat in London', *London Journal*, 29 (2004), 1–16.

Taylor, E. G. R., *Tudor Geography 1485–1583* (London, 1930).

Taylor, Katie, 'Vernacular geometry: between the senses and reason', *BSHM Bulletin*, 26/3 (2011), 147–59.

Thomas-Stanford, Charles, *Early Editions of Euclid's Elements* (London, 1926); reprinted with additional plates by Alan Wofsy Fine Arts (San Francisco, 1977).

Thorndike, Lynn, *The Sphere of Sacrobosco and its Commentators* (Chicago, 1949).

Trow, Martin, 'The business of learning', *The Times Higher Education Supplement*, 8 October 1993, p. 20.

Turing, A. M., 'On computable numbers, with an application to the Entscheidungsproblem', *Proceedings of the London Mathematical Society*, 2nd ser., 42 (1936), 230–65.

Turner, Gerard L'E., *Elizabethan Instrument Makers: The Origins of the London Trade in Precision Instrument Making* (Oxford, 2000).

Ullmann, W., 'On the influence of Geoffrey of Monmouth in English history', in C. Bauer, L. Boehm and M. Müller (eds), *Speculum historiale: Geschichte im Spiegel von Geschichtsschreibung und Geschichtsdeutung* (Munich, 1965), pp. 257–63.

Underhill, Evelyn (ed.), *The Cloud of Unknowing* (1922) *www.sacred-texts.com/chr/cou/cou09.htm*.

Van Egmond, Warren, *Practical Mathematics in the Italian Renaissance: A Catalog of Italian Abbacus Manuscripts and Printed Books to 1600* (Florence, 1980).

Van Egmond, Warren, 'Abbacus arithmetic', in I. Grattan-Guinness (ed.), *Companion Encyclopedia of the History and Philosophy of the Mathematical Sciences* (London, 1994), vol. 1, pp. 200–9.

Venn, John, and J. A. Venn, *Alumni Cantabrigienses*, Part 1: *From the Earliest Times to 1751*, vol. 3 (Cambridge, 1924).

Walker, Garthine, 'Modernization', in *idem* (ed.), *Writing Early Modern History* (London, 2005), pp. 25–48.

Wallis, Faith, 'Signs and senses: diagnosis and prognosis in early medieval pulse and urine texts', *Social History of Medicine*, 13 (2000), 265–78.

Wardley, Peter and Pauline White, 'The Arithmeticke Project: a collaborative research study of the diffusion of Hindu–Arabic numerals', *Journal of the Family and Community Historical Research Society*, 6/1 (2003), 1–17.

Watts, John, 'Public or plebs: the changing meaning of the "Commons", 1381–1549', in Huw Pryce and John Watts (eds), *Power and Identity in the Middle Ages: Essays in Memory of Rees Davies* (Oxford, 2007), pp. 242–60.

Wear, Andrew, *Knowledge and Practice in English Medicine, 1550–1680* (Cambridge, 2000).

Webster, Charles (ed.), *Health, Medicine and Mortality in the Sixteenth Century* (Cambridge, 1979).

White, Lynn, Jr, 'Jacopo Aconcio as an engineer', *American Historical Review*, 72/2 (1967), 425–44.

Whittle, Jane and Elizabeth Grifffiths, *Consumption and Gender in the Early Seventeenth-Century Household: The World of Alice Le Strange* (Oxford, 2012).

Willan, T. S., *The Early History of the Russia Company, 1553–1603* (Manchester, 1956).

Williams, Glanmor, *The Welsh and their Religion: Historical Essays* (Cardiff, 1991).

Williams, Glanmor, *Wales and the Reformation* (Cardiff, 1997).

Williams, Glanmor and Robert Owen Jones (eds), *The Celts and the Renaissance: Tradition and Innovation. Proceedings of the Eighth International Congress of Celtic Studies 1987* (Cardiff, 1990).

Williams, Gwyn A., *Madoc: The Making of a Myth* (London, 1979).

Williams, Ifor, 'The Computus fragment', *Bulletin of the Board of Celtic Studies*, 3/4 (1927), 245–72.

Williams, Ifor and Thomas Roberts, *Cywyddau Dafydd ap Gwilym a'i Gyfoeswyr* (Cardiff, 1935).

Willan, T. S., *The Early History of the Russia Company, 1553–1603* (Manchester, 1956).

Williams, J. E. Caerwyn, 'Gutun Owain', in A. O. H. Jarman, G. R. Hughes and D. Johnston (eds), *A Guide to Welsh Literature 1282–c.1550*, vol. 2 (Cardiff, 1997), pp. 240–55.

Williams, Jack, 'Mathematics and the alloying of coinage 1202–1700: Part I', *Annals of Science*, 52/3 (1995), 213–34 and 'Mathematics and the alloying of coinage 1202–1700: Part II', ibid, 235–63.

Williams, Jack, *Robert Recorde: Tudor Polymath, Expositor and Practitioner of Computation* (London, 2011).

Wilson, Jean, 'Queen Elizabeth I as Urania', *Journal of the Warburg and Courtauld Institutes*, 69 (2006), 151–73.

Wittkower, Rudolf, *Architectural Principles in the Age of Humanism*, Studies of the Warburg Institute, 19, 3rd edn (London, 1962).

Wood, Anthony à, *Athenae Oxonienses: An Exact History of All the Writers and Bishops who have had their Education in the University of Oxford*, vol. 1 (new edn with additions by Philip Bliss, London, 1813).

Young, Sidney, *The Annals of the Barber-Surgeons of London* (London, 1890; reproduced by AMS Press, New York, 1978).

Zilsel, Edgar, 'The sociological roots of science', *American Journal of Sociology*, 47/4 (1942), 544–62.

Index